T0073625

LAB
HOPPING

LAB HOPPING

A Journey to Find
India's Women in Science

AASHIMA DOGRA
NANDITA JAYARAJ

PENGUIN
VIKING
An imprint of Penguin Random House

VIKING

USA | Canada | UK | Ireland | Australia
New Zealand | India | South Africa | China

Viking is part of the Penguin Random House group of companies
whose addresses can be found at global.penguinrandomhouse.com

Published by Penguin Random House India Pvt. Ltd
4th Floor, Capital Tower 1, MG Road,
Gurugram 122 002, Haryana, India

Penguin
Random House
India

First published in Viking by Penguin Random House India 2023

10 9 8 7 6 5 4 3 2 1

Lab Hopping: A Journey to Find India's Women in Science is based on the
personal lived experiences of the authors and the individuals they have interviewed.
Some names and identifying details have been changed to protect the privacy of
the individuals quoted in this book. The views and opinions expressed in this
book are the authors' and the interviewees' own, the incidents related to their
personal experiences are as have been narrated by them, and the interviews
reproduced herein are as have been narrated to the authors and have been edited
and condensed for clarity. Some of the subject matter of this book relates to the
experiences of women in Indian science, including issues like gender discrimination,
caste, sexual harassment and sexual assault. As such, reader discretion is advised.
Penguin Random House India is in no way liable for the same.

ISBN 9780670090990

Typeset in Minion Pro by Manipal Technologies Limited, Manipal
Printed at Thomson Press India Ltd, New Delhi

www.penguin.co.in

Contents

Part III
Noticing Patterns

Part IV
Our Science Culture Must Change

Foreword

I was once introduced by a senior biologist at a lecture I was giving, with the comment that I was an *unusual* woman. Unusual because I happened to be good at mathematics.

It was a moment that left me momentarily bewildered, vacillating between gratitude and offence. In the end, the audience answered on my behalf, gently admonishing the old gentleman with a ripple of mocking laughter. I can still remember the look of confusion on his face. To his mind, he was just paying me a compliment. He didn't realize that, in so doing, he was insulting all women.

It was a minor error on his part, but it pointed to a widespread problem. There remain lingering doubts in the hearts of many people that women can *really* do science, or at least do it as well as men. Thankfully, it is a fallacy that's undermined with every passing year as more women reach the peaks of their departments, join scientific academies and win Nobel Prizes—slowly undoing long legacies of deliberate exclusion. When, years ago, I had the pleasure of being on a panel on sexism in science with Biocon's Kiran Mazumdar Shaw (who was also interviewed for this book), she struck me as the most self-assured person I had ever met. She took no nonsense.

Time is proving the sexist dinosaurs wrong. Eventually, all they will be left with is the sound of mocking laughter ringing in their ears.

But there's a long way to go until that day, and the wait is made longer by our failures to fix the deep-rooted problems that keep women from progressing in academia as quickly as men. The public might expect science these days to be a bastion of meritocracy and collegiate fairness—a world in which everyone is equal—but of course, as anyone who has actually worked in a lab will know, nothing could be further from the truth. Ego, status, money, honour, politics and prejudice—they all affect the work of researchers everywhere in the world. Scientists are human, after all. They're not gods (as much as some of them would like us to believe they are).

Science is, in fact, as human an endeavour as there could be. At the heart of a true researcher, whatever their other qualities, are passion and curiosity—those most beautiful and basic of human virtues, observed even in little babies. And it's those virtues that radiate from so many of the stories in *Lab Hopping*. Beyond the cruel reality of having to work in environments that don't always welcome them, the people in this book are chasing their intellectual dreams, uncovering fundamental truths for the good of society, and sometimes even sacrificing aspects of their own personal lives—forgoing motherhood or time with their partners, even disguising themselves temporarily as men—to do the work they love.

There have been many books written about women in science in the West, but too few about women and non-binary scientists in India. It's thrilling to finally have their stories in their own words and unvarnished, so we can also see beneath the glossier public image.

But of course, as inspirational as it is to hear from those who have had exciting careers, it is equally dispiriting to read their stories of discrimination and struggle. It's a bizarre fact of academia that people feel punished for doing their jobs or for even having those jobs at all. Where there should be gratitude for the work they do, there is too often instead abuse, sexism and harassment.

Nandita Jayaraj and Aashima Dogra describe themselves as 'sisters under the patriarchy'. This is perhaps the same umbrella under which women, transgender and non-binary scientists in India also find themselves. Routine indignities and casual slights are the shared everyday burden of the underestimated. We carry them like mental scars, hoping for eventual justice. For some, the abuse extends into the physical as well as the mental.

When I was a teenager, our schoolteachers told us that if ever we were confronted by a perverted stranger in public deliberately exposing himself in front of us—known in those days as a 'flasher'—the best thing we could do was to laugh loudly and walk away, ridiculing him and taking away any sense of power he might have. I can't explain why that episode came to mind while I was reading this book. I suspect it's because the need of some men to exercise control over the women they work with is so desperate as to be embarrassing. It demeans the men, not the women.

The damage that discrimination inflicts on science cuts in every direction. It is of the self-sabotaging kind. Good research requires humility, openness and fairness. It demands that scientists embrace expertise and talent wherever it is, even among those they personally consider inferior to themselves.

As Aashima and Nandita so beautifully explain in these pages, mediocrity in Indian science is sadly a product of its own shameful lack of diversity. If research in India is ever going to break barriers, if it is ever to be genuinely world-leading as it aspires to be, then it needs to get over its prejudices for its own sake—if not for the millions of women, non-binary and transgender people, and those from marginalized castes, religious minorities and lower socioeconomic backgrounds, who deserve to have happy, fulfilling careers without fear of unfairly privileged men peering pathetically over their shoulders.

Nobody should be made to feel unusual for being a scientist.

January 2023   Angela Saini
Award-winning science journalist and author

Part I

How It Began

1

Introduction

Our journey to meet Indian women in science began as the year 2015 ended. As science communicators and feminists, we were starving for science stories with women in them. Even with backgrounds as science reporters, we struggled to name any active Indian scientists who were not cis men. Yes, there were those women from the Indian Space Research Organization (ISRO) in bright sarees we all saw in that famous photograph after the successful Mangalyaan launch in 2014—but who were they, what science do they do exactly? A desire to know Indian women in science grew within us. Instead of waiting around for somebody else to tell us, we set out to find out for ourselves.

Our experiences before we set off on the 'lab-hopping' journey set the context better.

We were both employed as part of the editorial team that produced a monthly science magazine for children from an office in Bengaluru. The work there was fulfilling in many ways, but there was a lingering feeling that we were being taken for a ride—a common experience among women who work in Indian offices dominated by men. The boss of the company—a serial start-up entrepreneur—ran the show like a classic patriarch. He hired us, a very young, talented and imaginative group of writers, artists and programmers to work under him. He played the role of a patient and progressive leader very well, making us believe that we were equal stakeholders; but it didn't take long for him to establish his autocracy. When there was resistance to his style of working, he reacted strongly. We were tasked with creating all the content with the art and tech teams, while he handled marketing and finances. We would put in the creative labour and he would bring in the numbers. All dealings with him were extremely gendered—whenever we campaigned for transparency and fairness, he protested with: 'Ladies, please.' When it became clear to him that the ladies were not as agreeable as he wished, he began sidelining us. Whenever a man was hired in the small company, regardless of the hierarchy in place, the newcomer would have special meetings with him that we knew nothing about. There was definitely a boys' club forming as seed money for the start-up arrived.

On 8 March 2015, the investors invited all the employees for a special lunch on the terrace of the Bengaluru office. We ate standard lunch meals there every day, so when we arrived on the terrace on special invitation, we were extra-curious. To mark International Women's Day, the terrace had been divided for two separated genders; the women on one side were treated with samosas and chips and the men got standard meals. When we returned to our desks, eyes rolled up into our heads, bellies and hearts half full with the empty gestures, we found presents on our desks: handbags with graphics of lipstick and shopping and other 'feminine' symbols on them. All the stereotyping was infuriating. We returned the bags to the HR team along with an email full of rage. The day ended with the two of us debating whether pepper spray would have been a more empowering gesture from the employers on the occasion of Women's Day, in the context of our space and time.

Perhaps, this episode was when we truly recognized each other as sisters under the patriarchy. We were close friends already but our relationship became more purposeful and collaborative from here on. The issues of the magazine we since made together show that. Often, illustration drafts to go along with our science-y comics, poems and stories featured the stereotypical vision of a scientist—an old wild-haired man in a lab coat, shaking test tubes. Now with each other's support and solidarity we had the resolve to boldly ask for re-draws so that the images in the magazine we produced weren't full of old eccentric boffins. By the end of 2015, the magazine with thousands of subscribers below the age of fifteen shut down. The boys' club had not managed to pull off the numbers, letting the young readers down. We wrote the last editor's note with tears in our eyes, determined not to fall prey to the boys' club again.

We then set off on a new journey to challenge the stereotypical image of a scientist. The two of us went in different directions: one went northwards and the other roamed labs in the southern states, but with the same mission—to find and write stories of Indian women in science we met while hopping through science labs.

The first ever lab we visited was in Kalimpong. A hilly town where Nepali and Tibetan tribes settled long ago, Kalimpong has historically been a hub for exchange with the East, given its strategic location between Bhutan, Sikkim, China and Nepal. Nathu La Pass, where India and China fought in the 1960s is not far. The people, sick of being sidelined by the West Bengal state authorities, have put up a fight for their own statehood of Gorkhaland, which really began in the times of British rule. Their demand for self-governance is often met with indifference and

the revolt has sometimes been bloody. This leaves the proud
people with an urgency to preserve their cultural identity—their
own and of their lands. Part of this identity is a famous fruit,
the Darjeeling mandarin that grows in the hilly orchards of the
Gorkhaland area. Sadly, the orchards where these mandarins grow
have been under attack from a virus for three decades. Fewer and
fewer mandarins are harvested and sold every year. As the hill
agronomy suffers, cultural identity is collateral damage. Many
farmers are forced to grow the more profitable large cardamom,
somewhat unwillingly, over the well-loved mandarins.

Lab-hopping, as we dubbed this exercise, began at the
principal's office of a local Kalimpong college. After giving us a
gist of the local history, the principal and the geography lecturer
directed us to another institute. Their own college was only an
undergraduate-level college, so there was hardly anything there
to qualify as scientific research, they said. Our best bet was the
'Virus Office'. The 'Virus Office' is the local name for the Indian
Agricultural Research Institute's Kalimpong station. A bit out
of town, the Virus Office overlooks a green valley; disease-free
saplings are grown for distribution, and workshops for local
farmers also take place here. In the building, only two scientists
were stationed. One of them, the head, made some calls and
introduced his junior, a young woman named Natasha Gurung,
a fruit scientist.[1]

Natasha comes from the neighbouring state of Sikkim,
where her family still lives, three hours up and down the
valley. She gave us a tour of her plant tissue culture lab
featuring many brightly coloured mandarins. After the tour,
we sat down in her office to speak about her contributions
in identifying the virus that had been creating havoc in
the area, her motivation for reinstating the famously sweet
Darjeeling mandarins as the local cash crop and her many

ideas for modernizing local hill agriculture, including citrus oil extraction from mandarin rinds.

During the first lab-hopping interview, besides us and Natasha, there were two others in the room. While speaking, Natasha often paused for addendums that talked highly of one of them: Husnara Sharma. Husnara had been working at the Virus Office for much longer than Natasha had, but as a laboratory assistant. It was from Husnara that Natasha had learnt the ropes at the Virus Office. Where Natasha fell short in answering our questions about scientific and institutional matters, Husnara filled in.

The fourth individual in the room was Aman, Husnara's four-year-old son. He insisted repeatedly for cartoon videos on YouTube on Natasha's office computer. This is where some prickly naivete on our part was about to be lifted. As science communicators, we were looking for 'empowering' stories of science featuring women. That could have been the depth of our motivations when the project began. And here was one, in front of us: a scientific laboratory with scientific women. The double burden that many women in science carry (or triple burden as it has been called by sociologists to factor in the pressures of science in addition to formal work) had dragged itself into the lab uninvited. Meeting Aman, a child dependent on his mother, in a scientific space was for us, at first, irksome. What's a child doing here? Doesn't he get in the way of all the science?

In contrast, Natasha and Husnara heard his demands patiently and complied. Clearly, Natasha greatly admired Husnara's skilful management of the 'triple burden'. She looked up to Husnara in every way. So where was our own indignation coming from? After some self-reflection, humility, observation, reading and numerous interviews with some incredible women, we came to adopt the feminist lens through which one needs to look at the realities of Indian women in science. We had to shed our self-congratulatory view of 'young independent females' and negotiate our entitled position of 'freedom' to travel anywhere unhindered, devoid of realities like family and children. Meeting Indian women in science brought us face to face with all that we had taken for granted for many years—'talents' acquired as caste- and class-privileged, city-bred women. Before the lab-hopping journey actually took place and the reporting began, our view of feminism had been coloured by our privileges. It took a while for us to realize that our small wins against the patriarchy, like managing to stay out all night, were incomparable to Husnara's triumph in bringing her son to work. Our struggles had been very different from the struggles of so many women trying to keep their jobs, getting hired in the first place or fighting to remove a sexual predator from their surroundings. Husnara, Aman, Natasha and some other early interviewees opened the door for us . . . the door into the ground realities of Indian women in science, where we found stories of success and struggle.

The door was opened, and in we went.

Over the years, we spent hundreds of hours scanning websites of universities, institutes, conferences, science academies and government departments looking for women. The job was easy enough, but not without disappointments. In our search, women scientists stuck out like buoys in an ocean of cis men who dominate the landscape of Indian science. On the homepages of institutional websites, there is often an image slider that flips through the latest newsmakers at the institution. We made it a point to wait at the homepage for a few moments to watch the image slider run its course. What greeted us most of the time were men shaking hands with other men, men cutting ribbons, men receiving prizes, men posing in front of the university signage or men emphatically delivering lectures. Women were hard to find on these digital faces of Indian science institutes. Of course, they do make an appearance to supply the candles and hold the trays when the lamp needs to be lit.

Offline, our experiences were similar. When we set off to a place where scientists are likely to be found, we often began with long walks through the hallways, keeping an eye out for female-sounding names and/or '(Ms)', '(Mrs)' '(Smt.)' after the title Dr on name plates on lab doors. More often than not, the doors seemed to say to us 'Bad luck. Keep moving.' Sometimes, we ended up in the offices of well-meaning men, talking with them about the missing women with whom they should be sharing corridors, and the institutional realities that ensure their absence. They often prodded us towards one or two invisible women in their surroundings. Many times, these conversations took place in solidarity with our mission to investigate the gender gap.

From the beginning, we made sure to resist the pressures that would sway us into meeting only an elite group who are thought of as scientists worthy to be written about or those stationed in heavily funded government institutes. We did not want a sanitized version of Indian science. We intended to see it for what it is through a feminist lens. We did not go looking for powerful Indian women to feature; our attempt was to hand the mic to all kinds of Indian women doing science and turn up the sound. All of them were women, most were cis-gendered and predominantly from upper castes. As we searched further, we also got in touch with transgender scientists and those from marginalized castes, classes and communities. We met them in their places of work, whether in laboratories, on the field, in hospitals, in their offices, conferences, meetings, classrooms, universities, colleges and sometimes in homes or cafes, spread across the country.

Lab-hopping took us from the biting realities of public health to awe-inspiring discoveries of particle physics, from super-specific fields like palynology (the study of pollen), to broad ones like phenomenology (the study of strange phenomena), from highly theoretical sciences of string theory to much more translatable sciences like construction engineering, from lab-based sciences like genetics to field-based ones like observational ecology, from pollution to nutrition, from art history to animation . . . We met number theory wizards, discoverers of new species, quantum computators, neuroscientists who study human emotions, some very precise mechanical engineers and some very vague seismologists. We met a condensed matter physicist who moonlights as an artist, an archaeometallurgist who is a classical dancer, and a chemist who likes to bake. We heard from those who have been tirelessly campaigning for rational thought and the end of superstition in Indian society, for big science experiments to be conducted in the country and also those shaking their fists at policymakers for lack of climate action, inequity and for regulation on cosmetic products. We talked to experts about their new ideas in advanced materials for digital storage, game-changing energy solutions and had conversations about new drug targets. Some were lucky enough to have expensive machinery costing crores in their labs, many sent or travelled with their samples to other labs. We met struggling technologists, dejected ecologists and doctors and teachers longing for research.

In some women scientists, we saw revolutionary feminists and in many others a nonchalance towards, sometimes even aggression against, intersectional movements in the country. We ended up meeting more biologists than we wanted to. And we struggled to meet any ISRO women involved in the popular space missions. We interviewed Anu Sabhlok, a social scientist who has been searching for new ways to define 'a nation' with her ethnographic study of labourers from Jharkhand who build the India–Tibet border roads every year. And also Ayusmati, a PhD student on her way to becoming an oceanographer by analysing ancient climate from cores extracted from the bottom of the Arabian Sea. One out of the two Ramadevis we met had just retired after decades of inventing gadgets for spacecrafts as an ISRO engineer and the other had recently returned from the US to reboot the veterinary department at Banaras Hindu University. Many of them shared eye-opening thoughts with our readers, like when ecologist Jis Sebastian said: 'I feel safer in the forest than a city' or when transgender activist and doctor Aqsa Shaikh reminisced on her difficult formative years saying, 'I have only brought them pride.' We heard nuanced takes like: 'I am performing my gender' and also defensive views like 'I am definitely not a bra-burning feminist.'

The scientists we met had beaten the odds stacked high against them. They made up the measly 15 per cent that constitute India's women scientists. Studies and surveys have shown that there are several intriguing aspects to the gender gap in science in India. We get into some of them in delicious detail over the course of this book. No doubt, the numbers behind the gender gap helped us make sense of what we were seeing on the ground, but something else was happening too. One story after another, we were narrativizing the gender gap with real stories of women in science in India. We hope through this book, our effort to reveal the complex tapestry bears fruit. This has always been the focus of lab-hopping. Taking this approach seems necessary considering that data on women in science that has been quantifying the gender gap for decades has only been met with inaction. The large number of surveys, headcounts and in-depth studies performed by a handful of scientists and sociologists in the country studying the gender gap in science have not translated to significant changes in policy.

Our interviews put real faces to the percentages and graphs that populate studies and surveys. There emerged a consensus that societal expectations of family and childcare were only one of the many factors behind the exodus of women out of Indian science. Equal and often greater threats to their scientific dreams were hostile workspaces and an academic track that is designed specifically for upper-caste cis-men scientists. Even among the women who do make it, an indiscriminate number come from socio-economically dominant castes, religions and classes.

A cursory evaluation of the situation today suggests that things have improved. The quantity of women-in-science discourse has mushroomed even within the relatively short time frame of seven years since our project began. On 11 February 2022, dozens of institutes around the country organized talks and panel discussions to celebrate International Day of Women and Girls in Science. For our team, comprising some of the only journalists to specialize in reporting on this issue, this day is the busiest of the year. Government officials joined the flurry of action with the Union Science Minister announcing three new programmes to benefit women scientists, and also launching a booklet of success stories. The following month, on International Women's Day, UNESCO, with the support of India's Department of Science and Technology (DST), launched a swanky coffee table book summarizing the situation of women in science in India. Titled *A Braided River*, the gender imbalance-centric book featured visuals from an international photojournalist and made for a disconcertingly feel-good read.[2] The same day, a college in Mysuru announced one day of period leave per month for its women staff.[3]

At the same time, schoolgirls around the country are being restricted from classrooms because of their choice to wear a hijab.[4] A transgender student is desperate for an opportunity to continue doing science after psychological abuse drove her out of college. A contractually employed university professor is being forced to let go of her hired research scholar because her prestigious government fellowship has not reached her for a year. A PhD student has just emerged from a hunger strike to protest against her casteist boss.[5] Another grad student is on the brink of quitting her lab, unable to turn a blind eye to the activities of her predatory Principal Investigators (PIs are the term for scientists who lead labs). Some months ago, a young and academically brilliant researcher at a shiny new IIT wrote 'I quit' on the wall of her room before taking her own life.[6]

The cognitive dissonance is deafening.

Dozens of reports striving to identify the reasons for and the ways out of the gender gap in Indian science have been published over the last two decades. No longer possible to ignore, 'Women in STEM [science, technology, engineering and mathematics]' has become a hot-button issue, prompting governments and corporations to show that they care. Yet, the level of their engagement is very inadequate. While they are all eager to pat themselves and each other on the back for how far we have (apparently) come, nobody wants to talk about the areas where we are collectively failing. In all the prevalent discourse about the gender gap, discrimination is rarely dwelled upon. This is very counterproductive. Because how long can we talk about the plight of the discriminated without confronting the fact that there are very real individuals and institutions *doing* the discriminating? How long can we harp on about under-representation without questioning the systems that enabled a few groups to become so over-represented?

'Lab-hopping' is our attempt to upturn some of these habits. It will show readers that we cannot hem and haw our way out of a problem that stems from centuries of patriarchal oppression. Six years of conversations with those struggling and thriving in the bias-filled world of science and journalistic investigations into the various aspects of the gender gap through the years have convinced us that only a more brave and radical approach can address the gender gap in an inclusive way.

This book depicts inequalities in Indian science through the true un-sugarcoated stories of hundreds of scientists we have met in our travels. We have tried our best to not be held back by the fear of hurt feelings and the backlash of those who stand to lose privileges. It has not been easy. With such a volatile political atmosphere, we are constantly haunted by the pressures to 'look at the bright side of things'. But somewhere along the way, we realized that we didn't need to try so hard. The love for science, the struggles and the unafraid voices in this book are precisely the bright side of things. They are our strongest pillars of hope.

2

About the Book

This book is a commitment to seeing diversity in Indian science. Diversity, as has been scientifically proven, is essential in achieving excellence in science and technology as it opens up a dialogue between a wide range of experiences and perspectives, thereby unlocking the true potential of science.[1] The mediocrity in Indian science, we contend, is a symptom of its lack of diversity.

And in our goal to invoke diversity, what follows is an account of our science media project 'Lab-hopping', in which we carried out extensive investigations and interviews focusing on gender and other inequities in Indian science. With only close to 15 per cent of faculty members in research institutes constituted by women, science in India, like most other fields, has a gender problem. However, an earnest mission for diversity must be driven by the voices and experiences of the marginalized—in this case women. Lab-hopping has been an attempt to shift the narratives in science from a male-centric to a more egalitarian one. Besides increasing the visibility of Indian women scientists, the project has discussed the nuances of the gender gap in Indian academia and weighs solutions to close it by collectivizing many marginalized voices from across the country. In over 300 interviews and reports, we found intriguing commonalities among their journeys, as well as varied challenges they face while doing science in small and big institutes.

The two of us started the 'Lab-hopping' project as a science media portal in 2016 to document the journeys of active women scientists in India. Though it began primarily as a science communication exercise to report the science happening in our backyards in an engaging way, we quickly realized that a focus on women in science called for something more. It became increasingly impossible to write stories of women in science without getting into the nuances of their experiences. Soon enough, we steered towards the feminist politics of women and other minorities engaging with the scientific method in the country. Over the years, we were able to collaborate with a growing community of freelance science communicators, among them students, researchers, journalists, artists, photographers, film-makers, podcast producers and technologists. With this evolving and highly collaborative approach, the posts on our website thelifeofscience.com opened up to the hard questions that came our way.

From having to navigate relatively easy questions like 'How do we choose who is worthy enough to profile?' we had to confront more fundamental ones like 'Who is a scientist?' and indeed 'Who is a woman?' The more self-aware we allowed ourselves to become, the more our own invisible biases presented themselves to us. We knew early on that by focusing on the newsmakers, we would only end up replicating the most dominant narratives of women in science in India that we wanted to challenge. So, we deliberately chose scientists at random, occupying all levels of the hierarchy and representing as many geographies as our budgets would allow us. We included the voices of PhD students and postdoctoral researchers who were hoping to establish themselves in the science world. As journalists who were part of an Indian media that was just opening up to including science in mainstream reportage, we grabbed the chance to write about as many fields of science as possible.

Deliberations with economists, sociologists and artists convinced us that research in the humanities and social sciences—often referred to as 'soft' sciences—could be just as rigorous in terms of scientific temperament and method; just as the natural sciences can involve as much creative temperament as arts and humanities. Besides, though it may seem that the so-called 'soft' sciences do not have a gender problem, a scan through lists of prominent experts from art, economics and other fields from our country challenges that notion. Along the way, we also came to realize the importance of the contributions of other people besides 'scientists' who are integral parts of the laboratory—some highly qualified, others with basic education and also those dusting the equipment and sweeping the lab floor. We felt that to present a true picture of Indian women in science, we need to incorporate the experiences of all women who are part of the research ecosystem into our field of vision.

We followed the slow journalism approach, the subculture that questions the need for speed in this digital age, to get at the core of the gender gap investigation. Each report platforming the voice of a different woman scientist added to the discussion. Between 2016 and 2022, we published over 150 in-depth profiles of women in science, and another 120 posts that ranged from personal essays to news reportages and opinion pieces. We also engaged with the topic via social media in ways that are harder to quantify. Soon enough, nuances of the gender gap in Indian science started to be recorded.

What do we mean by 'Indian woman'?

In this book, we are discussing 'women in science', a terminology affixed within the gender binary that conflates two concepts, biological sex and gender.[2] While gender is defined as a social construct that is an internal sense of self, sex is something that is biologically defined by external or internal body parts. It was only a few years into our project that we started to ask our interviewees the question 'Do you identify as a woman?' Most of our interviewees are cis women, meaning that they were assigned the sex female at birth. The word 'woman' in this book spans through these different spheres of womanhood, sex and gender and lands on identity. Since we wanted to take a snapshot specifically of Indian science, we only featured women working in India.

Who is a scientist?

Someone engaging with the scientific (and research) method employing hypothesis, experimentation, analysis and sharing the results with the community at large is generally considered a scientist. As mentioned earlier, in our quest, the subject of such a scientific pursuit has no boundaries. Therefore, we considered an anthropologist just as much of a scientist as a materials scientist. In India, the term scientist is usually reserved for the PIs, the lab leaders who hire PhD students and postdocs, in their team using funds they are able to raise. While lab-hopping, we did not always comply with this definition. Our studies indicated that the non-PIs—lecturers, research scholars, grad students, postdocs, facility managers and technicians—were also important players in the Indian scientific ecosystem because of the important roles they play as well as the fact that they often had similar trajectories as the PIs.

With this book, we attempt to story-tell our way as we go on to investigate the gender gap in Indian science today. We have started by setting the context for our readers; we already laid bare to you our intentions in the previous chapter, and in this chapter, we share some methodology-related notes we felt would be useful as you go along. Numbers are important and that's why we also include a chapter on what we mean when we say 'gender gap'. This part of the book will end with some historical context through an account of some of India's historical women in science—there are more of them than you think! With this, the actual 'Lab-hopping' begins. The second part of the book, titled 'What We Saw', depicts different aspects of the gender gap as observed during our travels across the country; in the third part of the book, 'Noticing Patterns', we unpack some of the recurrent themes of bias that cropped up during our reporting of realities on the ground between 2016 and 2022. In the fourth and final part, 'Our Science Culture Must Change', the two of us encircle ideas from the feminist movement in Indian science that is determined to close the gender gap and transform science as we know by including all marginalized identities as Indian science progresses forward.

To protect our sources, we have used fake names (marked with * in first mentions) and also fictionalized the details of where they work in some places.

For the sake of bringing continuity to our lab-hopping tales, we will be self-referencing by using the word 'we' when talking about either one of the lab-hoppers. In reality, the two of us hardly ever lab-hopped together but since this book conjoins both our realizations, we choose to speak as a unit.

3

The Gender Gap

The gender gap in Indian science resembles a pyramid. More women are at the bottom of the layer of science education, from where their numbers start thinning out as we go further to the level of scientist jobs. The pyramid peaks at the top where women in leadership of science institutes are largely missing.

Only 29.3 per cent of the world's science researchers are women, reported a 2019 UNESCO factsheet. Bringing up the rear of the list of countries, there is India, where only 13.9 per cent of scientists are women. This gap is worse than all 37 Asian countries accounted for in the report, save one, Nepal, which comes last in the list with only 7.8 per cent women scientists.[1] In 2022, responding to a question asked in the Rajya Sabha, the then Science and Technology Minister Jitendra Singh said that women constituted 16.6 per cent of active scientists in the country in 2018.[2] Details of how this number was calculated were not made clear. In any case, we can safely assume the proportion of women scientists in the country is close to 15 per cent.[3]

The number of girls and women enrolled in science education and training has been rising slowly but steadily in the last decade.[4] According to statistics from the World Bank in 2021, women make up 43 per cent of postgraduates in science in India.[5] The numbers gathered from the All India Survey of Higher Education (AISHE) conducted by the Ministry of Education reveal that in 2010, 33 per cent of science PhDs were awarded to women, and around 2018, this number grew beyond 40 per cent for the first time.[6] The numbers of Indian women in science up till PhD and postdoc levels are truly encouraging. They contrast the gender gap trends seen in more developed countries where attrition of women in science begins at the lower college level. According to official reports and various media, in 2018–19, India ranked 3rd in terms of the number of PhDs awarded in Science and Engineering after the USA and China.[7] A *Hindu Business Line* report in 2020 even went as far as declaring: 'We produce the highest number of female graduates in the world—40 per cent of STEM graduates are women.'[8] However, this is where India's supremacy in the matter ends. Beyond PhD and postdoc levels, a widening gender gap in India persists.[9] As we tracked this gender gap in the last decades through different aspects, with studies, data and interviews, we grew sceptical of the hope for steady and gradual growth in women's participation in Indian science. We came across several examples where the numbers of women scientists reduced or stagnated in the last decades. The national science academies that are meant to take up the mantle of advancing science are included in the sources of this disappointment.

India boasts of three national science academies, the Indian Academy of Sciences (IASc) situated in Bengaluru, the Indian National Science Academy in Delhi (INSA) and the National Academy of Sciences, India (NASI), also known as the Allahabad academy. A fellowship at one of these scientific societies that lasts a lifetime is considered a stepping stone to the higher positions in Indian science. In 2015, the same year the lab-hopping seeds were germinating, 9.83 per cent of the IASc fellows were women. As of March 2022, seven years later, this number had slipped to 9.6 per cent according to data available on the academy website. Compared to the growth seen in the previous decades, since 2015, the rise in women's participation in the academies has been sluggish. Over in Delhi, INSA had 6 per cent women fellows in 2015 and no women took governing posts at the academy. The Allahabad academy's (NASI's) record in recent years is no better, without any considerable jump from the 8.1 per cent women fellows recorded in 2015. Even though a few women scientists have taken up top positions in the academies, in the absence of statistics or an intensifying dialogue on diversity from the top, we cannot say that the gender gap shows signs of closing in our academies.[10]

Statistics that show the disappointing reality of meagre support for women's research are evident from the reports of the DST. It is the nodal department under India's Ministry of Science and Technology that coordinates and funds much of the scientific activities in the country. In its annual report from 2010, the department acknowledged supporting 1,334 women PIs, the official term for lead scientist positions in our institutes. Four years later, the DST supported a lesser number of 1,301 women PIs. In the year 2019, the 'Directory of Extramural Research and Development Projects' released by the DST reports another instance of reducing support for women:

Women participation as Principal Investigators of Extramural Projects was 28 per cent in comparison to 31 per cent during 2017–18. These female PIs have taken up 26 per cent of the total 4616 projects this year in comparison to 29 per cent of total projects 4137 during the previous year of 2017–18.[11]

Another report, 'Research & Development Statistics at a Glance', consolidated funding support from various central science and technology agencies, including the Defence Research and Development Organisation (DRDO), the Indian Council of Agricultural Research (ICAR), Indian Council of Medical Research (ICMR) and DST. It also makes it evident in its many graphs and figures that the central agencies' backing of projects led by women crossed the 30 per cent mark only in the years 2010 and 2014, and in the latest available statistics from 2016, it had dipped to 24 per cent, the same number seen 10 years before in 2006.[12]

The gender gap at the science jobs level prompts the question, why aren't Indian science institutions hiring and supporting women scientists? Is it because they are scientifically unproductive? This would be true if the biases against women we heard while lab-hopping are to be believed. We cover these in depth in a later part of the book. These include claims that women are not interested in science or lack scientific aptitude. On the contrary, a study by researchers at India's largest R&D organization, the Council of Scientific and Industrial Research (CSIR), done as early as 1999 had the following conclusion:

The study contradicts the common belief that male scientists are more productive than their female counterparts using statistical test, since no difference between the productivity distributions of male and female scientists had been observed at the overall agency level of the Council of Scientific and Industrial Research. The same results has been observed in the data at the group of laboratories level, characterized by the subjects of physical, biological, and engineering sciences [sic].[13]

Even so, almost two decades later in 2018, a count of women scientists in India found that 36 lab-based CSIR institutes employed 4600 active scientists, of which only about 18.5 per cent were women.[14]

Besides funding, another way to gauge the activities of Indian women in science is to look at the top scientific disciplines in the country. Some years ago, the DST asked Elsevier—the largest global scientific publisher infamous for paywalling publicly funded knowledge—to run the numbers on India's research output between the years 2009 and 2013. This output was measured and compared to other countries by very exotic calculations that deal with how many research publications Indians publish and how many times these papers are referenced by the global community and various other factors. The report found engineering to be the most 'impactful' field and also states that 'relative to its overall output, India publishes particularly large proportions of its excellent papers in Engineering and Computer Science'.[15] These are the fields where women's participation is the lowest. Engineering is the field with the least women scientists. According to BiasWatch India, an independent group keeping tabs on stream-specific science 'manels' and corresponding women faculty base rates in Indian science, the base rate of engineering women faculty was the lowest in its list at 9.2 per cent and computer science was at 12.2 per cent.[16]

The absence of women from these high output fields is also seen at the student level. In the 2017–18 AISHE, it is reported that 91 per cent of India's PhDs in mechanical engineering (the largest engineering field) were men. The latest AISHE report that we accessed was released in 2021, and it reports a similar reality indicating no considerable change. In its summary the AISHE report states: 'At Ph.D. level, maximum number of students are enrolled in the Engineering and Technology stream followed by Science (chemistry, physics and biology).'[17] This mega report from the Ministry of Education is largely limited to the enrollments in streams of higher education, so we do not get deep insights into the proportions and counts of scientists employed in research institutions. For this we have to look elsewhere.

One extensive report that looks into and also comments on employed women scientists is the 'Country Report on Status of Women in Science and Technology from India', published by the Association of Academies and Societies of Sciences in Asia (AASSA) in 2015. Particle physicist and gender equality crusader Rohini Godbole was one of the two authors of this study. It found that women scientists have a larger representation in medical sciences, biological sciences and mathematics.[18] These are the fields where Indian science is less impactful on the world stage, at least according to Elsevier.[19]

Rohini's report makes three very clear conclusions. They are:

1. There is significant participation of women studying and teaching science in schools and undergraduate colleges.
2. However, this is not true when it comes to pursuing scientific research as a career.
3. The percentage of women faculty and students in science and engineering decreases with the perceived high status of the institution as well as with increasing position of authority within the hierarchy.

The third and final conclusion is critical from the feminist lens as it points to the lopsided power balance favouring men that cements the patriarchal culture that we see in Indian science. It explains where power is concentrated and systematically kept away from the bulk of scientific women in the country. The institutions that the study perceives to have a high status include well-funded institutions enjoying the most resources. These include organizations under the CSIR, DRDO, DST, the Department of Atomic Energy (DAE), Institutions of Eminence such as the Indian Institute of Science (IISc), Bengaluru, and Centres of Excellence such as the Tata Institutes of Fundamental Research (TIFRs) in Hyderabad, Mumbai and Bengaluru, and a few major universities like Delhi University. These are institutions perceived as 'high status', presumably due to large funding injections and the autonomy they are offered that frees them up from the checks and balances that other institutes governed by the University Grants Commission (UGC) have to adhere to. For example, in 2018, IISc was named one of the five public institutes to enjoy the status of an 'Institute of Eminence (IoE)'. The IoE tag could be compared to the US's elite Ivy League. As per a media report, these institutes would become part of a '"free" process without any checks from UGC' and 'get additional funding of up to Rs 1,000 crore' as well as 'complete academic and administrative autonomy, and complete financial autonomy to spend the resources raised and allocated'.[20] These 'Eminent' and 'Excellent' institutes are where India is most invested in science education and research. They are also the least inclusive academic spaces in the country. The IISc, for example, had only an 8.6 per cent share of women scientists among its faculty as reported in the 2021 AISHE report. This pattern of 'low inclusion–more resources' also extends to the gender gap among science students.[21]

According to Rohini's study and other sources, the largest numbers of women scientists are found at the bottom level of the pyramid of power and resources—most at state universities, fewer in central universities, tapering further in government research institutes. Women are hardly seen in institutes of national importance and the four academies; they are also largely absent among the list of leaders, science awardees and those holding prestigious fellowships.[22]

The gender gap goes deeper than just the attrition of women along the academic pipeline. These fault lines in the fields of science go beyond gender in a country like ours. Caste, being the biggest social reality in India, bears heavy on inequality in Indian science. Gender- and caste-segregated data are not available but thanks to a RTI application from 2021, it was revealed that 88.3 per cent of all the teaching faculty in 23 IITs belong to the dominant castes and only 0.6 per cent belonged to Scheduled Tribes, 3.1 per cent are from Scheduled Castes and only 8.1 per cent were from Other Backward Castes.[23]

In a 2021 lab-hopping article on the casteism in Indian science, a source was quoted as saying: 'IIT [name redacted] is like an *agraharam*', correlating the culture at one of the IITs to the neighbourhood within a village exclusively reserved for Brahmin priests and their families.[24] As we write, three academics associated with the IITs from so-called lower castes are fighting against atrocities that took place inside campuses. One of them is Vipin Veetil, former assistant professor at IIT Madras. In his letter to the National Commission of Backward Castes detailing the dire conditions of his dismissal from IIT Madras and demanding a fair fact-finding investigation, Vipin Veetil writes how he was hounded out of the institute when he made a complaint of harassment by four faculty there:[25]

By issuing various notices, one of which was pasted at my residential quarter in the presence of two witnesses, IIT

Madras's Administration made it very difficult for me to live and work through its intimidation tactics. I had to resign from the Institute due to the mental harassment meted out by the Administration.

My story is illustrative of the experience of faculty members from the 'lower' castes at India's 'higher' educational institutions.

This discordance is presumably even starker in elite 'specialist' research institutes, which curiously are not required to follow caste-based reservation policies, than in universities and smaller institutions where caste-based reservation policies must be followed.[26] Caste dominance is seen equally when looking at women scientists alone. It is then no surprise that the prominent voices of women in science in media and public outreach we hear come from the small fraction of women in top research institutes and mostly come from upper-caste and upper-class backgrounds. Dalit researchers with whom we discussed the gender gap told us that their survival within Indian academia has been contingent on the silencing of their own narratives and relegating conversations around their caste identity outside of the science ecosystem.

The many facets of the gender gap pervade through all fractures of our social fabric. Bringing an end to it is a hard ask, so we will instead point out that the statistics described above hardly depict a 'gender' gap. All of the above-mentioned data and reports are deeply rooted in the gender binary. The wrong assumption that people can be divided into the two genders male and female makes such an approach to discussing gender in science extremely error prone. But this is what our national dialogue on gender in science is currently based on. Gender is not the basis of these studies and reports, sex is. The conflation of biological sex (which is in itself not a binary) with gender has only served to invisibilize many intersex and transgender people in Indian science. The call for gender-affirming data collection has been loud from the trans-inclusive dialogue in Indian science, which is surely forming thanks to the few out and proud people we met on our lab-hopping journeys.

The only reason our records, statistics, conversations and world views have been stuck in the sex binary is because dividing humans in two categories is convenient for dominant groups. That comes at the heavy cost of exclusion.

4

Legendary Women in Science

Women in science is a topic that is in vogue as we are writing. It wasn't the case when we started on our lab-hopping journeys with the aim to fill the gender gap in our own reporting of Indian science. The scholarship and popular commentary on 'women in science' we referenced along the way are peppered with words like 'bias', 'invisible' and 'hidden'. These point to the fact that there always have been women in science but these lives and their stories were invisibilized and/or forgotten. In historical accounts of women in science across the world, we are met with examples of the few resolute women, photographed in black and white, facing the camera as sternly as they fought for their right to a life of science.

The most famous woman in science in the world is Marie Curie. She is the veritable mascot of the popular discussion on women in science, even among young science enthusiasts in India. Whenever we got a chance, during reporting or at events at schools and college, we asked the question: Can you name a famous woman scientist? Almost always we got the answer: Marie Curie. The story of the first woman Nobel laureate who won not one but two Nobel prizes is legendary.

Marie Curie's story radiates through history books as it starts us off by painting our first image of a woman scientist. It is a picture of her life of science highlighted by her discovery of the first known radioactive elements, radium and polonium, findings that would change the understanding of physics and chemistry forever. Along with success, struggle also characterizes her legend. Marie was a true pathfinder. Despite the extremely patriarchal and xenophobic attacks against her throughout her life, she stayed on and even thrived in science. Marie's passion survived through her two daughters, one of whom carried on her scientific work and also won a Nobel. Marie's own life came

Marie Curie

to a tragic poetic end from sickness caused due to radiation exposure. Her inspiring story, however, lives on.[1]

One of the most famous Indian women in modern science is Anna Mani. At the time that Anna was born in 1918, only around 1 per cent of Indian women were literate. Anna herself was born into many privileges, as part of a well-to-do Syrian Christian family in Kerala, headed by her civil engineer father. Anna grew up surrounded by books. The legend goes that a young Anna once rejected her family's customary gift of a diamond—the same material she would many years later perform experiments on. She asked instead for books. Her family was only happy to comply, an attitude that led Anna to pursue research and become India's pioneering meteorologist.

Anna's mastery in physical systems was forged at a lab at the IISc in Bengaluru, which was run by a more famous historic Indian scientist named C. V. Raman. Raman is a figure with many achievements under his belt, including the theory of Raman Scattering—the discovery that light changes energy and its wavelength when passing through a material. He also debunked the simplistic reasoning that the reflection of the sky causes the blue colour of the oceans and helped explain it correctly with the Raman effect of molecular diffraction in water molecules. His contributions to science earned him a Nobel Prize that he greatly coveted, a Bharat Ratna and many other honours. To this day, India marks the day he discovered the Raman Effect as National Science Day. Raman might have been a good scientist, but he was no revolutionary. It is worth mentioning his infamy in India's women in science movement, which is owed to, first, his reluctance in admitting women students to IISc, Bengaluru, when he was Director and then to his sexist policy of segregating spaces in his lab according to the gender of his students. In all his glory as a world-renowned Indian scientist, Raman could not secure for his student Anna the PhD degree she deserved. She was denied it due to the technicality that she did not possess an official master's degree. The lack of institutional recognition hit us hard as we went through her thesis at the library at the Raman Research Institute in Bengaluru in 2020. Although we have since learnt that being awarded the PhD or not was no matter that concerned Anna.[2]

From Raman's lab at the IISc, Anna went to Imperial College London, after which she returned to an independent India to set up new science institutions for the new nation. Towards the end of her career, Anna found a home at the Raman Research Institute (RRI)—a spin-off of Raman's legacy in Indian physics that was set up a stone's throw from the IISc.

Today many accomplished Indian physicists working here carry out research at the forefront of Indian science, including the search for superconductivity at room temperature, quantum communications and astrophysics. According to a 2019 gender count by Indian women physicists, less than 25 per cent of the faculty at RRI were women, and it has never had a woman as its head.[3]

Anna Mani on the day of her farewell at Raman Research Institute, 1980

During our visit there to deliver the 4th Anna Mani lecture organized by the Astronomical Society of India (ASI), we interacted with scientists at the RRI, some of whom were happy to share stories of Anna from the time she walked the same halls and apparently suffered no fools. In an interview in 1993 with science historian Abha Sur, a year before Anna suffered a paralysing attack, she was recorded as saying: 'What is this hoopla over women in science?' followed by 'It must be getting very difficult for women to do science these days. We had no such problems in our time.' This statement seen in the context of issues discussed in this book testifies to the privilege of an upper-class woman, the only kind that had relatively easy access to a life in science.[4]

Sadly, the discourse around acclaimed science figures does not make any significant attempts to carry forward the important work of making education, and by extension science, uniformly accessible to all Indians. As a result, today a scientific career might be a possibility for all, but the path is strewn with obstacles for most people. It is not surprising then that though Brahmins constitute only 5 per cent of India's population, they constitute a disproportionate part of the scientist population in India. All three Nobel Prize–winning scientists from India—C. V. Raman, Subramaniam Chandrasekhar and Venkataraman Ramakrishnan—were Brahmins. A deeper look at the backgrounds of even the women who manage to carry on in the scientific track reveals that they belong to socioeconomically advantaged groups.

The most legendary examples of women in science we have today can be seen as inspirational role models, but they are not reformative. Examples of famous figures that dominate the history of Indian science such as C. V. Raman, and even women scientists like Anna Mani, did not contribute to opening up science for all Indians. That credit goes to social reformers and feminists living much before their times. Two of them are Savitribai Phule and Fatima Sheikh. Thanks to the unrelenting efforts of anti-caste activists in the country, Savitribai has been recognized as the mother of modern education in the country. With her husband Jyotirao, Savitribai set up the first schools in the country open to all backgrounds of students, including Brahmin girls and 'lower'-caste children, previously disallowed learning. History has recorded that Savitribai carried a second sari on her way to school every morning in central Pune. It was an everyday occurrence to have dung and stones thrown at her for corrupting the minds of young girls and backward classes with education. Nevertheless, a defiant Savitribai was prepared as she marched on to bring us where we are today. With Savitribai was Fatima Sheikh, the first Muslim woman teacher. Not much is known about her, but we do know that they worked collectively and fearlessly towards universal education that crossed rigid social boundaries of caste and religion in the early 1800s. The education Savitribai and Fatima provided universally included the study of science and social studies, which was a lot more than that offered by the Indian gurukul, where the study of Brahminical scriptures took centre stage.[5]

The Indian woman's journey in modern science started in medicine. Until the nineteenth century, only men were being trained in Western medicine by the colonial regime. The indigenous '*vaids*' or traditional medics were also mostly male. Having women be examined and treated by a man who was not the husband was unacceptable to Indian society fully in the grips of patriarchy. Women across the country paid for this with their lives. In a time like this, Pandita Ramabai, another feminist reformer, minced no words when she spoke against the cultural patriarchy in Indian society. She wrote and spoke strongly on the

A stamp released by India Post in Savitribai Phule's honour

Courtesy: India Post, Government of India

plight of Indian women, and demanded the British provide medical education for Indian women. In 1882, she called on the Hunter Commission, which had been set up by the British Government of India, to look into the matter. She told the commission:[6]

> In ninety-nine cases out of a hundred the educated men of this country are opposed to female education and the proper position of women. If they observe the slightest fault, they magnify the grain of mustard-seed into a mountain, and try to ruin the character of a woman.

The plight of Indian subjects made the British Raj uncomfortable, prompting Lady Harriet Dufferin, wife of the then Viceroy, to launch in 1885 the Dufferin Fund, which would support women medical missionaries who had been arriving in the colony for a few years to provide medical treatment to Indian women. Part of this fund was to be spent to train Indian women in

Pandita Ramabai

Courtesy: Public domain, Wikimedia commons

Western medicine to spawn generations of Indian female medics to treat female colonial patients. As Geraldine Forbes, a scholar on colonial India, notes in her essay 'No "Science" for Lady Doctors', the key issue behind letting women pursue medical studies 'was gender, not science'.[7] The public health situation of Indian women was a disaster. And those in power felt the tension building up between the need to educate women and the reluctance of the patriarchy that operated. When medical institutions began to let in women students, the backlash from various groups was severe and included objections by the likes of the Council of Medical Colleges, the Office of Public Instruction and also different societies active in those times. Rhetoric to impress upon the authorities the unsuitability of women in education utilized falsehoods like women's intellectual deficits and absurd claims that there is no demand for education from Indian women. Neelam Kumar, in her important *Reader on Women in Science* describes education for Indian women and girls in the colonial times as a 'battlefield'.[8] Around 1880, amid the heated debates that are part of women's history around the world, the University of Calcutta, the University of Bombay and the University of Madras, along with the London University (now called University College London) became the first places of higher education to admit women students on equal terms as men.[9]

Part II

What We Saw

5

Women on Top

At the dawn of the twentieth century, Ida Scudder, the daughter of an American medical missionary living in south India, was moved by divine intervention to establish the Christian Medical College (CMC) in Vellore, Tamil Nadu. The story goes that one night, three Indian men knocked on Ida's door, one after another, asking for her urgent help in handling the labour complications of their respective wives. Ida noted, when she retold the story years later, that one of these men was a Brahmin, another was Muslim and the third a 'high'-caste Hindu. She had to turn down all these pleas for help due to her lack of medical training but suggested that her father, who was a doctor, go with them instead. Having a man medically examine or treat their wives' bodies was unacceptable to these three husbands. In the morning, the news came that all three young women had passed away. To Ida, this was a cruel yet definite sign from God that she needed to do something. She immediately left for America, trained there as a medical doctor and returned to Vellore to lay the foundations of what would become the CMC in 1900.[1] Until the year of India's independence, CMC remained a women's medical college.

CMC has emerged as one of the biggest tertiary hospitals in the country. Rooted in Christian values of service to the underprivileged, even today, it remains a hub for modern science-based community health services. The institute moves medical mountains every day. Professionals from local hospitals around the country receive their training from the important healthcare node that is CMC Vellore. One might even say, north India has AIIMS and the south has CMC. Ranked as one of the top places to study medicine, CMC has been contributing to India's medical research and pushing its boundaries for over a century—a commendable feat by a minority-run institution. The institute has maintained top quality while retaining the values of community medicine that were part of Ida's vision from the beginning. It has also possibly seen more women heads of departments than most other premier research institutes in the country.

Along the heavily trafficked main street of Vellore, the large landscape of the CMC building complex is crowned by a blue church-like facade. The Christian cross studs this crown. Thousands of people regardless of their religion pass through the gates seeking medical care. Dodging ambulances, patients and white-coated doctors, we passed the emergency wing, where several stretchers are parked. Walking past the consulting rooms with large crowds huddling in waiting areas, we arrived at the back of the hospital campus. There are no patients in these buildings, but on almost every floor an open window awaits medical samples arriving from consultations, research projects and clinical trials being conducted all around the country. Next door are the offices of CMC researchers who double up as teachers training the next generation of medics. Among those we met there are Priscilla Rupali, who has been working on controlling the spread of antimicrobial resistance in the country; Priya Abraham, who followed the spread of hepatitis virus and went on to head the National Institute of Virology in Pune during the COVID-19 pandemic; and Vrisha Madhuri, a paediatric-orthopaedic surgeon who is credited with designing a brace to help children born with clubfoot. Also among them is Sitara Ajjampur, a parasitologist leading the Indian side of a large international clinical trial attempting to break the transmission of intestinal worms like roundworms, whipworms and hookworms through community-wide deworming drives.

Seven of the ten women scientists we met at CMC shared with us very proudly that they had been mentored by one of the most accomplished scientists in the country. Anna Pulimood, who was the Principal of CMC when we went lab-hopping there in 2019, said: 'One of my very close friends has reached the peak that any woman can reach as a scientist in the world.'

'Is that friend Gagandeep Kang?' we asked. 'Yes, that is Gagandeep,' she said.

Gagandeep Kang, called Cherry by friends and family, is a medical doctor turned scientist. Among contemporary Indian women scientists, she is someone who gets closest to 'top scientist' status. During the pandemic, amid the unprecedented levels of fear, chaos and pseudoscience that spewed from all directions, she was a voice of science and reason, a calm in the storm. She appeared on numerous TV programmes, wrote many articles, co-authored a book on COVID-19 and participated in several scientific panels, day after day, to demystify the pandemic for us all, as an expert in infectious diseases. When not in the limelight, she was part of several government committees, including the one on vaccine research (her main area of expertise) that was mysteriously disbanded by the Indian Council of Medical Research (ICMR) in April 2020. She continues to offer her expertise at the top level 'as and when required'. And she does it all with the apparent ease of a synchronized swimmer.

When we met her at her office in 2019, she was at the tail-end of her time at the Translational Health Science and Technology Institute (THSTI) in Faridabad. In 2016, Cherry had taken on the Directorship of THSTI after 30 years of research on the Indian digestive tract. She spent most of her formative years at CMC in Vellore, where she retained her lab. CMC Vellore is where she returned after resigning as Director of THSTI in 2020.

For Cherry, it is routine to be running several large projects across the country (nine at the time of the 2019 interview), which includes conducting clinical trials for major vaccines. She is an important consultant to India's public health, a space where a lot needs to be done and she is not one to waste any time. Under her leadership, a scientific workforce of over 140 people, in the field and the lab, was responsible for successfully testing and launching a population-wide programme of vaccinating newborns against diarrhoea-causing Rotavirus, which is now part of India's Universal Immunization Programme. Cherry's research group conducted all phases of clinical trials and laboratory testing for about 15,000 children enrolled in different studies of rotavirus vaccines, two of which were successful in Phase 3 studies. The two successful vaccines, made by Bharat Biotech and the Serum Institute of India, are WHO-approved and are now being rolled out around the world. The post-licensing Phase 4 trials were underway in 2022 to test the effectiveness of the vaccine in about 35,000 children across India.[2]

The homegrown Rotavirus vaccine comes in two brands, Rotavac and Rotasiil, and costs the government one-twentieth of what the foreign brands charge in the private market. In our discussions with scientific officers within the government, the success of this Rotavirus vaccine often featured on top of their list of shining examples of India's recent scientific achievements. Cherry's lab in CMC has been testing the different versions of the vaccine prepared by multiple manufacturers for the Indian market and for studies done internationally. Standardized assays and reagents are also provided by her labs to support Rotavirus vaccine manufacturers in Brazil and China to evaluate the vaccines made there. Cherry's work over the last 20 years has provided evidence to show the need for the vaccines and prove that the vaccines work.

Her achievements are stacked tall. In 2019, she became the first woman scientist working in India to become a fellow of the Royal Society in London, the big daddy of the imperialist science enterprise that we now call modern science. The Royal Society fellowship that has been awarded since 1660 (to women only since 1945) is an achievement shared by the likes of Charles Darwin, Isaac Newton, Alan Turing, Srinivasa Ramanujan and Dorothy Hodgkin.

Those following Cherry's tracks have to climb high. Thankfully, help is at hand. From the top, Cherry has been laying out rope and harness for those who need it. Over the years, she has invested heavily in her students and others joining forces to work with her. Her mentees benefit from several opportunities to train in the skills required to be a good investigator. The goal for her, she told us, has been 'to see them be independent'. Among her many successes, scientific and otherwise, for Cherry, the most valuable has been her mentees. 'My definition of success is that if you can leave five people behind who work like you at the same level, your job is done,' she said. She then smiled before pointing out that she has managed this with not five, but eight people she today considers her protégés. 'And I'll be done in a couple of years,' she added very proudly in all her graceful charm.

One of these eight proteges is Sitara Ajjampur. When we asked Cherry to share the names of her mentees and collaborators, she responded over email with a long list, Sitara's name was followed with '-independent'. This means that she no longer works under Cherry but leads her own team in the work she receives grants for separately. We knew the hyphen symbolized immense pride for both mentee and mentor.

'Cherry has handed me more opportunities than I could deal with!' the jovial Sitara laughed as we settled in for a conversation at her office at CMC in Vellore. 'I have been very lucky in my career. For example, one cannot write a successful National Institute of Health (US) grant better than someone who has done grant-writing training AT the NIH,' she said, referring to one of these Cherry-sourced opportunities. Besides ensuring her proteges are at the right place at the right time, Sitara said her mentor's own track record is a big edge for anyone associated with her lab. 'Cherry has set up a fantastic lab, she is world-renowned, and has an established reputation. Because of this, the preconceived notion that collaborators have about us is always positive.' Such a privilege is shared by only a few scientists in the country.

Sitara casually met Cherry during her second year of MD at CMC. Over scientific discussions, the senior scientist invited her to join her lab as a PhD candidate researching parasites in human guts. Towards the end of her PhD in Cherry's lab, Sitara was pregnant with twin babies, 'which, you know, takes over your life'. She managed to submit her thesis, birth her kids and then defend her thesis in 2010. 'One of my sons was born with a cardiac problem; we had to wait for two years before he was old enough for an operation. During this time, my husband and I readjusted our work life to be with our son.'

During this delicate time, Cherry offered flexibility rarely seen in Indian science, making sure her protege had the right kind of support she needed. Today, Sitara is an independent PI and part of various international parasitology programmes and is also a happy mother of healthy twins. She is carrying out a trial with the support of the Government of India to collect evidence for community-wide deworming, which, according to her research, may work better than deworming only school children, as is currently done.

In November 2017, we attended the 'Women in Science—A Listening Session', a roundtable discussion organized by the Wellcome-DBT India Alliance in Delhi, to troubleshoot the gender gap in Indian science. At the meeting, Cherry appealed profusely to her peers to take mentorship seriously. 'In my experience, the women scientists of India are lacking good mentorship,' she said. 'It really does help to have a mentor. Mentors are useful both personally and professionally because they have been through what you are going through,' Cherry added with concern.

'To bridge the gender gap in India, why not just look at mentoring? Because to me that has been the most valuable approach.' Behind every great woman, are other great women. Cherry herself is no exception. 'The best thing that my own mentor ever told me on the personal front was how her grown-up child always complained: 'Mom, you were never there for any of my school events.' Yet, she said that she doesn't feel guilty because she knows she made it to enough of them. 'The daughter just doesn't remember!'

'Indian women, in particular, are guilted right, left and centre. We need to acknowledge ourselves that this guilting is going to happen and we need to be prepared to deal with it,' Cherry, mother of two adult sons, said as she addressed the room full of close to 40 leading Indian women scientists. The 'bad mother' trope is not the only one that haunts women with a heart for science. In labs across the country, other common tropes showered on the few women we have in power positions also include 'the bad boss', 'tough madam' and 'superwoman'. These tropes popped up frequently during our lab-hopping journeys. And indeed, as Cherry had cautioned, they do seem to have been internalized by people of all genders.

As we went on to profile more and more women scientists, we heard on the whisper vine, rumours of one of our interviewees 'throwing objects at her students' and another lady boss found to be very annoyingly conducting lab life by taking on the role of the helicopter parent. Sometimes, these claims came wrapped up in well-meaning fears that in our quest to publicly record stories of contemporary women in Indian science, we were running a risk of promoting those that didn't deserve it. Even when these anecdotal claims against women bosses were credible, we consistently found sweeping gender-based generalizations as we probed further. There was a pattern. No one ever said that 'Mr X, the male PI (lab leader), is a bad one'; but when talking about bad women PIs, we often heard references to their gender in the accusation. Bad experiences with bosses were being attributed to gender only when the subject was a woman. The sharing of unpleasant experiences with women at the top tended to vilify not just individuals who behaved badly but also their gender. The more we heard the claim that women make bad bosses because of this incident or the other, it became clearer that there was an underlying patriarchal construct at play, which held women to higher standards than everyone else. Especially when women made it to higher levels, their gender marginalization carried an extra burden of immaculacy, which, needless to say, men are exempt from.

Gendered tropes of bad women bosses play the role of showing women their place in a patriarchal society—at the bottom of social ranks, doomed to work the hardest and wield no power. We were reminded repeatedly that stereotypes working against women persist in our science culture today. The language our well-meaning informants were using was fitted with slurs reserved for women in power, namely 'bad mother', 'Queen Bee' and even 'dragon lady'.

As a scientist casually remarked over a conference dinner we attended in 2018: 'Women are capable of being as wicked as men.' But we met several postdocs and PhD students who believed male science bosses were more useful in challenging times compared to female ones. 'Never again will I choose to work under a woman,' we heard a PhD dropout from IISc say, describing a bitter PhD experience. 'If I have a problem, I go to male faculty, not female,' said a postdoc we met at TIFR professing their dislike for women bosses. These confessions prompted us to wonder if there is more to the 'Queen Bee' stereotype than internalized misogyny. We have a few ideas on this but first, let's establish that even though surveys and anecdotes that show preferences for male bosses are easy to find, this doesn't equate to 'women make bad bosses'. In fact, there are emerging accounts of how female leadership can be empowering for subordinates as well as their discipline/industry. The trend of preferring male bosses is only further proof that bias against women in science pervades all of society.

The Queen Bee phenomenon was described by US researchers as early as the 1970s to explain how women in senior positions distanced themselves from other women, especially when they were juniors and projected themselves as 'not like other women'.[3] The social psychologists that have discussed the Queen Bee phenomenon have been clear that this is a strategy employed by a marginalized group to evade their further marginalization. The fields of power that women leaders have to navigate are scattered with exclusion. And hence, being seen as any other woman only goes against their career progression, especially if their sights are set on the top of the ladder. Having lived through such experiences, when women emerge on top, they identify strongly with their own survival, persistence and hard work, which might lead them to distance themselves from younger women who might be struggling.

In a 2017 study titled 'The Queen Bee: A Myth? The Effect of Top-Level Female Leadership on Subordinate Females', Brazilian researchers Arvate et al. found a 'pro-female causal effect of female leadership in public organizations', meaning more women on top leads to more women in the organization. They also report that this effect is time-dependent; the longer a woman on top stays in power, the stronger her effect on improving the gender ratio. In their large-scale study of 8.3 million organizations, they write that the Queen Bee phenomenon is a response to gender inequality, a consequence of gender discrimination experienced in the workplace and not the cause of it.[4] Sadly, women in leadership positions are being directly incentivized to play by the rules put in place by men, thereby enabling male domination of institutions. The result is what can be called 'system justification' of the exclusion of women in science by women in science. This justification fans the fires of the widespread thinking that women can't survive in science unless they are extraordinary.

In the journeys of some women scientists we interviewed, the response to claims that women make unsatisfactory bosses strikes a contrast. Anju Bhasin has been the Founding Vice Chancellor (VC) at the Jammu Cluster University. When we met her in late 2019, Anju was and still is the first and only woman to ever head any of the 11 universities in Jammu and Kashmir.

She said, 'Women leaders are subjected to very different and a lot more expectations from juniors and peers alike. And it is not like we are heard or considered—I've seen this in government meetings too. If the few women that are in committees are not heard, then what is the point of having them there?'

This experience is shared by many of our other interviewees. Some professed that the shunning of women also extends into their labs, where they find it hard to get interested graduates and postdocs to do science with. After being the 'token woman' in many committees, Anju had had enough. To those claiming men make better science bosses than women, she says: 'Watch your bias!'

Anju's efforts in setting up the Jammu Cluster University became harder to realize with the abrogation of Article 370 from the Constitution furthering division in Jammu and Kashmir. Her attempts to promote dialogue and exchange between the students from Jammu and those from Kashmir proved to be an impossibility, she told us when we spoke to her again in October 2020. She had resigned as VC and left Jammu to continue her work on fabricating parts for particle accelerators at the European Organization for Nuclear Research (CERN) in Europe.

The few women on top we have in Indian science surpass expectations and survive the obstacle course in front of them on a daily basis, only to find themselves in a position considered somewhat odd for a woman to be in. The novelty in the famous picture of ISRO's women admin staff celebrating Mangalyaan entering Mars' orbit in 2014, and the reaction it got displayed how most people find the presence of Indian women in high sci-tech spaces slightly odd. Why else would the picture be so iconic? It was one of the most recent 'women on top' moments in the Indian consciousness, which also stuck out as an oddity at the same time. And popular science media from around the world documented how we all felt about it.

In a firstpost.com report, Sandip Roy, radio host, novelist and commentator wrote:[5]

> When we think about space, whenever we think about science, we think about Vikram Sarabhai or Homi Bhabha or Satish Dhawan. Serious men in suits. We do not think of women in brightly coloured silk saris, with a bit of gold on the borders, *pottus* on their forehead, and *gajras* in their hair whooping it up. We've seen pictures like that on Facebook but they are usually at Pongal or Navratri or wedding celebrations. But this was at the Indian Space Research Organisation.

In a BBC report titled 'India's Mars Mission: Picture That Spoke 1,000 Words', Pallava Bagla, India's veteran science journalist, is quoted as saying: 'The women were leading the applause when the good news arrived. They were celebrating more than men. Who said men are from Mars and women are from Venus?'[6] Such a reckoning seems to stem from the novelty of seeing Indian women in scientific spaces. It's worth noticing here that this 'odd' or 'novel' perception is imposed upon women in science from the outside. One may even say it is accrued by the male gaze.

The lived experiences of Indian women scientists that we have been recording present a different perspective, one that is their own. The women on top we interviewed in Indian science don't feel odd at all to be working in science, where they have earned their rightful place! One truly unabashedly tough woman who feels right at home in Indian science is Kiran Mazumdar Shaw. At the industry level, Kiran's iconic influence in Indian science is unquestionable. The role model in her signature pantsuit and bright silk scarves has inspired a generation of biotech entrepreneurs.

One of them, Supriya Kashikar, herself a businesswoman having cofounded GeNext Genomics, said in a media interview: 'Since class 12th, I used to admire Kiran Mazumdar Shaw because she as a scientist and woman brought the sweeping changes in our perception towards life sciences in India.'[7]

Kiran started Biocon, a biotech company in 1978, after taking the opportunity in extracting enzymes like papain from the tropical papayas that grow in the country. Today Biocon is a very successful company, with a revenue of over Rs 8000 crore in March 2022. It was the first company in the world to produce human insulin from a kind of yeast. With the largest number of diabetics anywhere, the market in India for the metabolic fixer hormone is humongous, although Biocon also has markets everywhere.[8]

'Are you aware you are an inspiration to many women in the country?' we asked Kiran, after managing to squeeze an interview into her busy schedule one day in 2019. 'Well, obviously, I would expect to be a role model, because of my work. If I'm not, then it will be disappointing,' she laughed in response. 'When I look at my own journey, I see that it is really my focus and my sense of purpose that has got me to where I am,' she said, sharing her formula to success as a woman on top in Indian science. She also added a message to her admirers and women everywhere: 'You can get to the top if you're in a mission mode.'

Among her academic peers, Kiran notices that the very successful ones are those who have global connections. In India, globally connected scientists are a rarity. 'Women scientists in India have even more limited contacts with scientific teams abroad,' Kiran pointed out: 'If you just want to be known in your own institute, it doesn't get you very far. To be successful, one needs to be globally networked in every possible way—whether it is through publications in journals across the world, whether it is through workshops or conferences, or getting newspapers to write about your work.'

In our own lab-hopping experiences, we hardly ever got the impression that women lacked the drive or the desire to hold positions of power. Instead, we heard them protesting that they were not being heard. Almost all the women we met agree— and often accept—that they, as women, have to work harder than their male counterparts. Some said women work twice as hard, and some others said they work eight times as hard as men. The social connections that Kiran suggests are required to be successful are often beyond the reach of women in science, especially those from marginalized communities. While most women rely largely on their 'human capital', meaning an investment of their own time, efforts and mental labours, their male counterparts enjoy their 'social capital' or connections to push them further. This lack of social capital might be the reason Indian women scientists are not seen commonly on the global platform, we suggested to Kiran. 'Call it what you want but I feel that women have to do a lot more to promote themselves in a more public way and understand what is happening.' Kiran rests her case.

The simplest way to get more women in science is to have more women on top. According to a World Bank review of studies from 2012, titled 'Gender Quotas and Female Leadership', that looked at gender quotas for top positions in India and Norway: 'The Indian evidence demonstrates that quotas increase female leadership and influences policy outcomes. In addition, rather than create a backlash against women, quotas can reduce gender discrimination in the long-term.'[9] This has been studied and proved extensively. Having more women on top not only ensures women's interests are considered in decision-making but also provides role models and mentors that, most of our interviewees insist, our scientific world desperately needs. As almost all top positions in science are held by an upper-caste old boys' club, their hold is unchecked and women and other marginalized people in leadership positions such as VCs, Secretaries of government science departments or Directors are not easy to find. As of May 2022, no woman has ever headed any of the Indian Institutes of Technology (IITs). Even at the lab-leadership level, the numbers are low. In fact, the percentage of women PIs in R&D projects reportedly dipped from 33 per cent to 24 per cent between 2013 and 2017.[10]

Our efforts to understand the gender dynamics in Indian science took us to a mega international conference of cell biologists, the ICCB which was hosted in India by the Centre for Cellular and Molecular Biology (CCMB) Hyderabad in 2018. The food hall looked as if it had to be stretched out to accommodate the large numbers of attendees, most of them presumably women PhD aspirants. Watching them pour into the hall was both awe-inspiring and unsettling. Almost all of them had come with hopes of being PIs one day, more than willing to make their contribution to the understanding of the fundamentals of life. The next Gagandeep and the next Kiran were probably among them. But after years of lab-hopping, we had understood the sad reality that only a very small percentage of them would make it unless there is a radical change over the next decade.

6

The Teacher–Scientists of India

For a scientist dreaming of a fulfilling life in academia, one of the country's 'Institutes of National Importance (INIs)' is the place to be. This includes elite institutes of higher education such as the IITs, All India Institutes of Medical Sciences (AIIMSs), Indian Institutes of Information Technology (IIITs) and Indian Institutes of Science Education and Research (IISERs). In 2016, it was announced that 20 higher educational institutions would further be recognized as Institutes of Eminence (IoE), an even more exclusive group that would be groomed by the Government of India to become world-class teaching and research institutions. The IoE, which seem to fancy themselves akin to the USA's Ivy League universities, comprises the IISc, a few of the IITs and some central and private universities. In the current scenario, a job at an IoE or INI comes with the promise of higher salaries, job security, societal status, and, above all, access to funding, infrastructure and social networks to do impactful research. These are also the places with the most pathetic overall gender ratios.

According to a national survey released in 2021,[1] only 19 per cent of the teaching staff at INIs are women. The stats get worse as we zoom in further. Around the same time, the IITs averaged 11.2 per cent of women among their faculty members and IISc was at the bottom with just 8.6 per cent. In terms of representation of women among faculty, India's public and private universities and colleges were doing much better at 38 per cent. Even when it comes to the enrollment of students, the situation is similar. The poorest gender ratio among students is seen in INIs (24.7 per cent compared to nearly 50 per cent in state and central universities).

It is clear that most of India's women in science are not in research laboratories but university and college classrooms. Unfortunately, these are the places that the government, by corollary, has deemed unimportant. And it shows.

As of 2022, 1027 universities were recognized by the UGC, the government body charged with coordinating higher education in India. Of these, some are run by the central government, some by state governments and others privately. These are the settings within which a large proportion of our women in science study, teach and research. Regardless of the breadth of universities spread across the country, it has become rare for noteworthy scientific research to emerge from them. Since the 2000s, the most popular scientists have hailed from one of the INIs or IoEs, which includes fewer than 200 institutions across the country. Central and state universities, on the other hand, have been making news more for their mismanagement rather than for academic accomplishments.

To infer from this that university faculty are simply not good enough scientists would be a mistake. Interestingly, the distancing of universities from research is a recent phenomenon. In the first 20 years since the inception of the prestigious science award Shanti Swarup Bhatnagar Prize in 1958, 40 awards went to university or college scientists, and 42 came in the next 20 years.[2] But in the last 20 years, only 15 Shanti Swarup Bhatnagar Prizes went to scientists at universities or colleges (*IISc and TIFR, which are technically deemed universities, were not included in these calculations). What caused this change and what does it mean?

The 'INI' status was introduced way back in the 1960s as part of the strategy to give select premier institutes more autonomy and, thereby, better chances at nurturing excellence. This move did not intend to sideline research happening at universities, but that is what eventually happened. Recently, certain policies have created a deep chasm between teaching and research. 'Leave research to the experts at INIs' is the message being thrust upon scientist–teachers at universities. As a result, newly minted scientists are all coveting faculty positions at INIs. The fortunate few who make it into this league have access to the resources they need to conduct world-class research. Naturally, more awards are now going to them. PhD students working in INIs, of which around 32 per cent are women, stand to hugely benefit from this exposure.

At universities, meanwhile, the situation is much more bleak. The purpose of universities and colleges has always been higher education, and for subjects like science, research is a big part of this. Research-based pedagogy, however, becomes an illusion when faculty are overworked with teaching and administration work. Not surprisingly, a lot of this labour is taken up by women working at universities. This leaves these 'teacher–scientists' very little time or incentive to conduct research. This is bad news for the nearly two lakh students doing their PhDs at universities, of which nearly 50 per cent are women. With limited exposure to quality research and mentorship, even those students with an inclination for research are under-equipped to compete with their resource-rich INI-trained peers for the best science jobs.

The saving grace for university students is the presence of talented and hardworking teacher–scientists who manage to stay inspired despite great odds. Among them are a significant number of highly qualified women scientists. These women became university teachers not necessarily because they preferred to teach over doing research—on the contrary, accepting a position at the university was the only viable way for them to stay connected with research. Swapna Mahapatra is one such professor at a state university in Odisha. We met her in 2017 and settled in for a chat in her large Head of Department office at Utkal University in Bhubaneswar.

Swapna, a high energy physicist, had joined Utkal University over 20 years ago. 'I love teaching, but I need more time for research. You can't do much of that in state universities, but sometimes to get one thing you have to give up something else.' Coming from a learned family in Odisha, she was privileged in that she did not have to struggle to get the education she wanted. She studied physics in local colleges in her hometown and began doctoral studies at the prestigious Institute of Physics in the capital city, Bhubaneswar. By this time, Swapna realized that research was her calling and it would need her full attention. She informed her family that marriage was not part of her plan; a PhD and higher studies came first.

As it turned out, Swapna met like-minded physicist Karmadeva Maharana during her PhD and married him the year she finished. The couple realized it would be impossible for them to raise a child in addition to pursuing research. She said, 'We made the conscious decision to not have children so we could focus on physics. Nowadays, people find that crazy—but you know, in those days, many of my friends too had similar ideas. The field is very demanding. I know that with a child I would not have been able to leave for Germany for my postdoc.'

The couple spent almost five years apart, while Swapna pursued postdoctoral research in famous laboratories in India and abroad. In the middle of a prestigious fellowship in Germany, she heard of an opening at Utkal University, where her husband was already a faculty member. A state university may not have been the natural first choice for an upcoming high energy physics researcher, but for Swapna, this was quite ideal. It would reunite her with her husband and enable her to care for their families, who needed the help. Teaching would take up a bulk of her time, but she made sure to keep her zeal for research alive. It helped that the German fellowship had a provision by which she could continue her work in Germany whenever she had the time. After she joined Utkal University, Swapna used the summer holidays to do this.

When we met Swapna in 2017, she was juggling multiple roles—Professor, Head of Department, Director of Research and Development and Dean of Science. She didn't have any airs that you'd expect from a powerful person though. Probably because in universities, leadership is more likely a double-edged sword. The titles may sound glamorous, but the workload is overwhelming. Holding all these positions didn't mean any less teaching, and it certainly didn't mean any more time for research. Swapna recalled some exceptionally tough years in the early 2010s when the bureaucratic red tape was complicating the hiring process at the university. As a result, there were only three faculty members scrambling to keep all bases covered. It took almost five whole years until new staff were recruited. Until then, Swapna and her two colleagues in the physics department had to endure practically impossible teaching hours. 'After back-to-back 90-minute classes, the brain will not work,' said the physicist recalling those frustrating days.

Such scenarios are fairly common in universities. So, it's not difficult to see why university teachers do not find the time for research. This is not only a hindrance to the teacher–scientist's professional ambitions, but it also compromises the quality of science education and exposure for students at the university training to become scientists themselves.

In 2010, the UGC announced that college teachers and university lecturers compulsorily had to conduct research and publish papers to climb up the academic ladder. While this might have sounded like a progressive idea at first, it soon became evident how disastrous it was to mandate overworked teachers to do research in grossly underfunded and under-resourced set-ups.[3] Sarbari Bhattacharya, a faculty member at Bangalore University, saw first-hand the problem with this situation: 'In state universities like mine, though there are multiple criteria for promotion, showing that you have done research seems to be the most influential one. For the sake of promotion, I have seen teachers choose to spend time on research by highly compromising their teaching. Some teachers don't even land up in class . . . they send their PhD students to take classes.'

Sarbari said that holding a professor position is typically required for a faculty member to be considered for powerful administrative positions. 'For those who have such ambitions, it makes sense for them to chase promotions,' she said. 'In general, I do think men tend to be more ambitious for these roles. And you will see that in most universities, these roles are held by men.'

Like Swapna, research is very close to Sarbari's heart; she too had trained in material physics in top institutes in India and Europe. She joined the University of Bangalore with the understanding that universities exist for the sake of training the next generation. However, the going has been tough. When we spoke with Sarbari, her department had not made any recruitments in the last 12 years, even though a large chunk of faculty members had retired during the period. She said, 'The sad thing I see here is that often [even after the heavy compromises on teaching], the research that comes out of here is not even really great. I would have no problem if it was, but often that's not the case. They just do enough for the requisite number of points to be promoted, and for this, the students are shortchanged. I don't think this is right.'

Sarbari's observations reflected reality because, during the period after the UGC's 2010 mandate, pay-to-publish or 'predatory' journals thrived. Publishing work in journals is how research is disseminated and science progresses. But 'number of publications' has devolved into a performance metric in academia. As the number of papers published turned into a parameter during faculty promotions, overburdened teacher–scientists had a strong incentive to publish in fake journals. Unscrupulous businesses exploited the situation and many scientists, it seems, were either duped or willing to dupe others just to stay in the promotion race. A global journalistic investigation revealed that the publishing rate of five major predatory publishers had tripled since 2013.[4] As part of these investigations, journalists from the *Indian Express* tracked down a slew of companies that churned out fake journals operating from Hyderabad. The companies charged 30 to 1800 US dollars as publishing fees. With this exposé, India found a prominent place on the world map of predatory publishing. And the blame fell on teacher–scientists at universities.

As research fraud mushroomed, there grew a widespread cry among top researchers in the country to stop forcing teachers to do research. They campaigned for universities and colleges to remove 'number of publications' as a criterion for faculty promotions. 'Let teachers teach' was the refrain. In 2017, the government buckled by announcing that research would no longer be a promotion criterion for college teachers. The dichotomy between teaching and research was reinstated.

Caught in between, but somehow shunted to the background of this to-and-fro, were the major stakeholders, the teacher–scientists themselves. Following the 2017 announcement, Renny Thomas, a sociologist who was at the time working at the University of Delhi's Jesus and Mary College, wrote in an opinion piece on TheWire.in:[5]

> How can one be a good teacher if she is not a good researcher? To deny their freedom to do research is to deny their freedom to teach that is supported by their research. It is to deny students the right to have good teachers as well.

Whether they were relieved that they no longer had to squeeze time for research, or disappointed that once again they were being told to leave the research to the 'experts' at INIs, university teachers were irritated with being told what to do and what not to do. Aruna Naorem, a faculty at Delhi University we first met in 2016, was one of the many who were displeased. She said, 'Teachers who are genuine and honest will do fair work, irrespective of the rules imposed on them and irrespective of where they work. I feel that the main culprit is the government trying to impose so many things on teachers on a day-to-day basis. This will create disparity among teachers. It's ruining the whole academic atmosphere, be it teaching or research.'

The 'Let teachers teach' campaign was demotivating for teacher–scientists in private universities too. Smitha Hegde, a teacher–scientist at a private autonomous college in Mangaluru, Karnataka, has fought many odds to keep her research alive. She said, 'The burden [of scientific fraud] will be relieved not by lowering the standards, but by employing more teachers, filling up the vacancies with fair, merit-based regular appointments, supporting undergraduate and postgraduate education in colleges with good administrative and infrastructure funds and having standard vigilance policies. I feel deeply saddened by the move [by the government to signal that teachers should stick to teaching], as it certainly is a step backwards in higher education. Decisions should be based on where we want to be and not a mere reaction to where we are now. I hope the college teachers will not face discrimination by funding agencies limiting their opportunities to grow.'

Smitha's angst is sincere. As one of the country's leading experts in pteridology—the study of ferns—she was in the middle of a project for the forest department when we first met in 2016. The Karnataka Government had enlisted her help to devise a strategy to control fern growth-related forest fires in the tiger-inhabited Kudremukh Forests. As happy as Smitha was to help the forests, she was just as grateful to receive the one lakh rupees that this project would pay her. Unfortunately, these trickles of funds were all that was keeping her lab running. Government funding for teacher–scientists like her is hard to come by, according to Smitha. PhD students at private colleges often do not have stipends or fellowships supporting them. Without funding, PIs have no money to pay PhD students. Smitha relied on projects like the Kudremukh one to purchase supplies for research.

Though private universities have the drawback of lower salaries and fewer perks (there are, of course, exceptions to this) and lesser funding from government agencies, they tend to do much better than government universities or INIs as far as the gender gap is concerned. This reflects both at the student (nearly 50 per cent of PhDs are women) and faculty levels (41 per cent of faculty are women). Smitha has a theory that partially explains this: 'An imbalance in sex ratio in any ecosystem points towards the emergence of a limiting factor. In this case, the limiting factor is remuneration. Women are more willing and able to accept jobs with lower pay. In spite of their legendary prowess for setting up good bargains for groceries and vegetables, they seldom bargain across interview panels/boards. They generally value the mere opportunity to work over income, perhaps, as theirs is probably a secondary income in the family and they do not experience the pressure of being the primary provider to their families. Thus, only one gender is thriving in the sector of education, and conditions of poor pay in schools and colleges continue to prevail.'

The large number and distribution of universities—be it central, state, deemed or private—make them automatically attractive to women scientists. This is because they tend to have more mobility restrictions than their male counterparts. However, once they begin working there, they have to get acclimatized to the less-than-ideal research environment. Often, this means teaching and administrative work consume their days and they have barely any time left to pursue their research interests. As if that wasn't challenging enough, they also have to navigate the politics and patriarchal attitudes that are much more explicit in universities than they are in elite institutions. Shanti Priya, an astronomer, had an especially turbulent experience when she was hired as Assistant Professor at Osmania University, a state public university in Hyderabad. The first-ever female recruit of the department, Shanti Priya was unusually young at the time, just 24 years old. She had just completed her MSc at the same university two years back. Shanti Priya was able to join as a faculty without a PhD because it was a 'backlog' post, a position that could not be filled for reasons unknown. Already nervous about a new job, she walked into an atmosphere of extreme hostility.

'The professors who were part of the faculty were very senior, they were the ones who taught me. They just could not accept that a 24-year-old *girl* had joined them and would now have to sit with them equally in all meetings. That was a problem,' she recalled matter-of-factly when we met in her office in 2017.

'I was given a room full of rats near the washroom. There was no separate ladies' washroom then, so it was a combined one. They said I would have to adjust there till a professor retires and vacates his room.'

Shanti Priya found herself having to ask for every basic requirement. 'When I asked for a computer, they said they never had computers in their time so I would have to work without one. But research cannot happen without a system! So I used to sit with the students in the computer lab.'

Shanti Priya knew that she needed to do a PhD to continue in the field, but this too ended up being a tedious pursuit. From 2007 to 2010, she ran around trying to convince any of the professors to be her guide. She shared, 'It took me three years to gain the trust of a professor I could work with. When I finally went for my first observation, I had to leave my six-month-old son behind for 10 days.'

We were further reminded of how toxic the work atmosphere at universities can get when we heard that a PhD scholar from a university in Kerala was being harassed by her doctoral committee for 'going after fame'. The committee was offended that she had agreed to be featured in a media article for her research work on sustainability. She was informed that the university had an unofficial ban on all communication with the media, and this violation could result in the rejection of her almost-completed thesis.

In this large-scale mistreatment and underutilization of scientists in universities, students become collateral damage. Being fewer in number, INIs are not as geographically widespread as universities, giving an automatic edge for a student in, say, Bengaluru and Pune, as compared to a student in northeast or central India. Within these 'blind spots', the female students are affected more, as their mobility—the freedom they have to move to faraway places for science—is typically much less.

Most of Utkal University's students are from Odisha or nearby West Bengal, and according to Swapna Mahapatra, extremely poor and hence unable to afford expensive engineering or medical exam coaching. She explained,' They come here because of their interest in science and because they know this is the best department in the university. People here publish in journals with good impact factors and the students rarely remain unemployed. Many join as PhD candidates in top research institutes and become lecturers, school teachers or faculty members in places all around the world. We motivate them and mentor them, it's part of our job. Good research is being done here, but I would not say it is comparable to, say, the IITs because the facilities and staff strength are much more there.'

In his speech at the Indian Science Congress 2019, Prime Minister Narendra Modi pointed out, 'The vast majority of students in the country go to state universities, where quality of research really needs to improve.'[6] He is not wrong—our universities, within which a large proportion of our women researchers study and teach, need help. But then it does not make sense to propagate the argument that universities have their place (teaching) and research institutes have their place (research). This point of view does a disservice to the Indian scientific workforce as well as students hoping to become part of the science community. After all, the way academia is structured is not fair or transparent enough to guarantee that the best minds for research are hired at INIs, while those more suited to teaching end up in colleges and universities.

In 2020, the Ministry of Human Resources Development shared some statistics which showed that the dropout rate in higher education was lower in elite institutes such as IITs and Indian Institutes of Management (IIMs) than in other institutes of higher education.[7] No surprises there. While INIs and IoEs boast of supercomputers, air-conditioned lecture halls, manicured gardens and comfortable housing, public universities often have to fight for basic needs such as women's toilets and internet connectivity. Between this, the arbitrary rules, unsupportive seniors, promotion politics, long teaching hours, administrative work and battling outside perceptions, it's no wonder that students and teachers at universities are discontented.

7

Rebel or Support

A lingering curiosity for us as journalists investigating gender gap realities was this: if so many women in science drop out before they join the scientific workforce, then what sets apart the ones who make it? We had some early insight into this during our 2016 visit to the historic Banaras Hindu University (BHU). When we met Kavita Shah, she had just been promoted to the post of Director of the Institute of Environment and Sustainable Development at the university. The biotechnologist could barely restrain her joy as she spoke with us about her life in science. During the next few hours in her presence, she not only impressed us with her research feats but also hit us with this 'truth bomb': 'There are two distinct kinds of women who can go very far in research, those who have support from their families and those who rebel.'

The astuteness of Kavita's statement didn't strike us immediately but dawned on us gradually as we progressed in our lab-hopping journey. Research is different from most professions because the day-to-day nature of the job comes with no rule book. No fixed number of years put towards a PhD or amount of postdoctoral experience guarantees one a place as a scientist. When starting out in her lab, a scientist is expected to don the hats of an HR person, civil engineer, accountant, counsellor and many more. Once her lab is set up, the next challenge is to be noticed by the research community and keep a lookout for opportunities. This is not easy to do in an (upper-caste) man's world. She needs to walk the extra mile, and then—if she is from any other marginalized group—walk a few miles more. Almost all the scientists we met had to make difficult compromises along their pursuit of a successful career, unless they were lucky enough to have support. Kavita comes from the camp of women scientists who have benefitted from unrelenting support from family and peers. In fact, most researchers we interviewed come from this 'support' camp.

For the woman scientist, 'support' takes on different avatars. For Jis Sebastian, a young Kerala-based conservation ecologist, support manifested in the form of nurturing parents. Though from a small village in the state, Jis's parents brought her up with the agency to dream big. At the age of 20, the young woman took her first ever train ride to join the prestigious Forest Research Institute in Dehradun. Since then, she has spent years living in remote jungles and befriending locals as she studies wild flora and fauna. At a relatively young age, Jis gained considerable recognition for her work as a crusader for conservation, sustainable tourism and ecofeminism. This is an unconventional lifestyle for a young woman in India, and it may have been impossible if she did not have the support of her family.

While a progressive-minded family is a big edge for a young woman aspiring to be a scientist, even more helpful is social capital, i.e., access to influential networks and, consequently, opportunities. In India, this has a strong correlation to caste, and this is apparent from the fact that almost all renowned scientists today—from any gender—are upper caste. Those growing up in such privileged backgrounds are able to receive the best of education and access to professional advice, as well as contacts that make their academic paths smoother than it is for others. They are empowered to take the socially unconventional decisions that a career in science sometimes demands. For women, the most rigid convention is arguably marriage, or more specifically marriage at the 'right time'. Breaking out of this convention is simpler for women fortunate enough to belong to an elite society that is less explicit in their judgement.

Jyotsna Dhawan is a prominent name in cell biology in India today. She is also refreshingly unshy to acknowledge her privileges. She grew up on the lush campus of IISc in Bengaluru where her father, aerospace pioneer Satish Dhawan, was Director. Jyotsna was encouraged to live a 'life of ideas', to take her own decisions, make mistakes and pick herself up. Before embarking upon an academic career, she ruminated for quite a few years about what she wanted to spend her life doing. When she decided to tie the knot, Jyotsna was 35. 'I had my fears about being able to handle family and work life, maybe that's why I married very late,' she said when we visited her laboratory at CCMB, Hyderabad, in 2017.

While such decisions are becoming increasingly common among women in urban India, the degree of rebellion and stress underlying them differs from circumstance to circumstance. For Jyotsna, it was relatively smooth sailing. She shared, 'We were, in a sense, protected from society because our family ethos that 'we had to do what we have to do' was so strong. This is a huge advantage because it made me mentally prepared. I've been really lucky that way. Things that slowed me down were all internal—my indecisions, my lack of confidence.'

'I've been lucky that way.'

This is possibly the most common sentiment we come across in our interviews. We hear it not just from privileged scientists, but also from those who come from less affluent backgrounds. As we met scientists across a wide range of geographies and socioeconomic backgrounds, we noticed something interesting and unexpected. For women, caste and class privileges do not always translate into more freedom. Upper-caste, well-off families are frequently just as restrictive with their daughters, sometimes even more. When we met the newly announced director of ISRO's much-celebrated human spaceflight programme at the headquarters in 2018, we found out a bit about her early life.[1] V. R. Lalithambika was a gifted child in a family of engineers in Kerala, but despite her social advantages, she wasn't exactly free. After completing a bachelor's degree, the electrical engineer was pressured by her grandmother to stay back in a local college so that she could get married and start a family. It did not matter that she had just aced an exam that qualified her to take up a master's course at the prestigious IIT Madras. Her marriage was promptly fixed and Lalithambika enrolled for an MTech in the local college. She was stoic as she related her story; there was no obvious regret in her tone.

As it turned out, things worked out for Lalithambika's career in spite of the limitations placed upon her mobility. Being close to home may even have turned into an advantage as she only had to take 41 days off to deliver her child during her master's. Eventually, she earned a position at ISRO's Vikram Sarabhai Space Centre, which is located in her home city of Thiruvananthapuram. Having access to childcare support from her family enabled her to live up to her potential at her workplace, and rise through the ranks to her current position.

Being pressured to compromise on education for the sake of marriage is a common obstacle in the life of a young woman in India, one that often spells doom for her research career. A woman with a PhD is frequently considered 'too educated' in social settings where men are expected to be higher qualified than their wives. Kannal,* a mathematician from Tirunelveli, told us how her controversial choice to pursue doctoral studies elicited disdainful comments such as 'Who will search for a boy for you now?' In such a scenario, higher education itself turns into an act of rebellion. It is illuminating to observe the various ways in which scientists have navigated through these social confines. Kannal fought to be allowed to leave for Karnataka to pursue her PhD at a prominent institute there; when we met her in 2017, she was busy with research, determined to achieve her goal of becoming a math teacher.

While such gutsy stories are rightly appreciated, it must also be stressed that rebellion is not a safe choice for many. Often, the choice to conform is a mode of self-preservation. Indian children, especially girls, are brought up to think of elders as gods, so the thought of acting against their wishes can be mentally distressing. Rebelling may very well mean complete withdrawal of support, and when 'family honour' is an issue, there may even be violent consequences. A senior scientist, Shravanthi,* spoke to us about being threatened with violence by her elder brother, the patriarch of the family. He refused to accept her choice to marry her lover, a fellow scientist who belonged to a different caste. She was banished from her family for over a decade and went through tremendous emotional turmoil. When we met her in 2019, she was the head of her department at the university and expressed no regrets. She was happy that things were cordial once again with her family.

Rebelling doesn't automatically imply a shot in the arm for one's career. Priya's* parents were very supportive of her when it came to education but when she expressed her desire to marry her senior, who comes from a different religious community, they severed ties with her. In the years that followed, the couple faced great difficulties raising their children, one of whom was born with a serious health condition. Without any social support system available to them, Priya had to drop out of science to keep her family afloat.

In spite of the possible complications, women in science are increasingly choosing to defy societal norms of marriage and children. More often than not, the gamble pays off. When we met Mamta Rani in 2016, she had one arm in a cast, but that didn't slow her down as she juggled tasks adeptly. She was busy setting up the Central University of Rajasthan's computer science department. Mamta's parents too were very enthusiastic about educating her. But soon enough, it became evident that Mamta's expectations for herself exceeded those of her parents. 'Being from a large family in a village, they wanted me to get a job at a safe place, close to home in Uttarakhand,' she said. Mamta refused to compromise. 'If I did this, I wouldn't have reached here.' She was 39 when we met her, and content with the life she's made for herself, believing that marriage will happen when and if it has to.

On the same campus as Mamta, we also met Hemlata Manglani, an economist who studies how self-help groups function in north India. Hemlata also faced intense pressure to marry from extended family members. With time and her rising credentials came respite, as she shared, 'From the beginning, I was a topper at my school and college, as well as a national-level debater. The awards and trophies I won spoke for me. Later on, even my grandmother realized that I was meant to go on studying. I'm still not married. I am extremely happy with this. And now, so is my family.'

Hemlata also emphasized the role of her mother in supporting her when no one else did. 'From the first standard to the fifth standard, my mother taught us. She had studied only till the third standard but she was keen for us to go on. Because of her support, I am the most educated in our community. No one has gone this far, neither girl nor boy.'

Instances of small and big rebellions are widespread in the stories of women in science in India. And not all of them are against societal obstacles. In the course of our travels, we came across numerous women, young and old, who had to rebel against bigoted institutional structures to claim their space in Indian science. They objected to sexism in the hiring committee, demanded justice from a sexual harasser, lobbied for a daycare centre . . . the examples are countless, and many of them will be discussed in upcoming chapters of this book. Each time we had such an encounter, Kavita's statement about needing to be a rebel to stay in science rang truer.

The trouble is, not all people are intrinsically rebellious. As Rekha Singhal, a researcher we met at the Indian Institute of Forest Management in Bhopal, pointed out, 'We discourage girls from asking questions. We tell them *Bahut savaal puchti ho*' [Don't ask questions], '*Bina jaan pehchanwalon se baat mat karo*' [Don't network], '*Jaisa kaha hai vaisa karo*' [Do as asked]. In this way, Indian girls are discouraged to think scientifically.'

When we are conditioned for years to suppress any rebellious streak we have, the idea of displeasing authorities can be painful and stress-inducing. We often wondered what gave some women the fortitude to do so. Was there a set of favourable conditions that combined to awaken the dormant rebel within them? What might go on in the mind of a young woman in science while choosing whether or not to rebel?

Jyotsna Dhawan believes that one powerful catalyst is a love for science. In her decades of guiding young researchers, she observed that the decision to commit to science usually comes to a youngster when they understand that they are good at it. 'The stage where something they predicted actually happens in experimental research—that's a powerful feeling. Then they begin to realize that a sustained effort at research is worthwhile. After that, it's almost like a drug,' she suggested.

For young Archita, this moment came when she joined a top institute as a PhD candidate in 2017. She was born and brought up in a conservative Brahmin family in the same city as her institute in Uttar Pradesh. Her mother was a homemaker and her father was an office superintendent at a local college. Archita never dreamt of becoming a researcher despite being outstanding at academics throughout her school and college life. She said, 'When I became a gold medallist in my MSc, then I knew I wanted to do something with my life. I was brought up to take up a job like teaching—something that would allow me to handle all the responsibilities expected of a girl. But coming here changed my way of thinking. Here, I saw many women doing fantastic work, devoting their whole life to it. Science, it is a celebration, a lifestyle. I was inspired by people finding interesting questions and ways to answer them.'

Having never left home and being sent to all-girls schools up till then, it was not easy for Archita's family to digest the fact that their daughter was leaving home to pursue research. She said, frustrated, 'My father and uncle did not agree with my decision. They think that research, having no definite time frame, will take me away from them. They are afraid I will want to leave home, live abroad . . . They don't understand research.'

Fortunately, Archita had the backing of her mother, who insisted that only education could give her the respect, honour and everything else that she could want.

With the reluctant permission of her family, Archita was strongly advised to choose a woman as her PhD guide. She did, and she has never regretted it.

'Ma'am is so polite to me, so supportive. I think she understands me and all aspects of my life. In the lab, she often comes over to my desk to find out how I'm doing, personally and professionally. I know now that in science there is no difference whether your guide is male or female, but it helps in the society that we are living in. My mother and father are happy that my guide is a woman.'

Archita's love for research consolidated over her first two years as a PhD student, but that passion alone will not ensure an escape from the infamous 'leaky pipeline'. When we met her, the pressure to agree to an arranged marriage was looming over her head, and Archita admitted to worrying if she had the strength to bear the price of defiance. 'There are times I have thought of giving up on my studies, but this is all I have. All that matters to me is my research,' she declared in earnest.

We first met Archita in 2018 as a bright-eyed and eager grad student who was part of the organization team of a scientific meeting being held at her institute. During dinner one night, we found ourselves at a table with Archita and a senior woman biologist who was one of her idols. Over the course of our conversations, the biologist seemingly sensed that Archita needed a pep talk. She advised the young researcher that it might be time to consider flying out of her home nest. She even brought up the possibility of Archita coming over to her lab for a few months to collaborate and experience an academic atmosphere away from the control of her family.

We followed up with Archita some months later and we found her full of hope. She called her interaction with the senior biologist 'the best conversation of my life':

'I returned to my room feeling like the confidence she put in my head and soul was going to change my life. She said you can do anything if you want to. Since then, I have been thinking of becoming a scientist like her. I hope that in the next few years I will have something to be proud of. I know now that I can get support when I want it. I'm not alone in this world.'

This particular scientific meeting was a memorable one for us as lab-hoppers. We got a chance to witness the solidarity among the women in developmental biology, not just over dinner table conversations, but also in the auditorium, when the few women at the meeting, such as stem cell expert Jyotsna Dhawan and neuroscientists Shubha Tole as well as Jonaki Sen, held the mic to engage in lively discussions of each other's work. This struck us as a deliberate contrast to the typically male-dominated discussions at most conferences. We recognized it as a subtle rebellion in its own right.

For transgender and non-binary persons pursuing science, the difficulties are amplified. It's not just a matter of overcoming one big obstacle like marriage or overprotective parents, rather, for them, every step involves a rebellion. The ignorance among most people of what it even means to be transgender and their exclusion from public spheres ensures that there is little support from biological families of trans persons who are 'out' and virtually have no mentors to show them the way in academia.

Grace Banu is an engineer by profession but a transgender activist by necessity. Her activism enabled her to study or lead the kind of life she wanted to. In an interview, Grace recounted how she was discriminated against in school for being a Dalit person.[2] She sat separately from the rest and had different timings. The persecution was so severe that Grace left the school. Meanwhile, she also left her childhood home for a trans commune where she found her chosen family. Here, Grace was encouraged to resume education. Though she could not become a doctor like she wanted to, Grace cracked the competitive entrance exams and joined an engineering college. Despite no scholarship support and a daily commute of five hours, Grace persisted. When she found that she could not afford the examination fees, Grace persuaded the college principal to change the system to make it easier for people like her. Thanks to her efforts, the university issued a notice that transgender students were exempted from paying the examination fees from then on.

Her rebellions continued even after she became a software engineer. When Grace's daughter Tharika was applying to medical colleges, the duo were dismayed to see no category for transgender persons. Having no other choice, they struck off 'Male' and 'Female' and wrote 'Transgender'—only to find out later that Tharika's application was rejected. There was no way, it seemed, that a transgender person could apply to medical college. Together, the mother and daughter filed a case at the High Court and secured a moral and legal victory when it was ruled that Tharika, being a transgender person with equal rights, would be given admission to the college. With her collective, Trans Rights Now, Grace routinely enables other transgender persons to access their rights to education and employment. Yet, each victory is a bittersweet feeling. She awaits a day when every transgender person will not have to go to court to get justice.

Maybe Kavita is right and there are only two kinds of women who can excel in research—those with support and the rebellious. She was speaking from gendered experiences of her own as well as those of her students, who fell back despite their love for science. In this scenario of privilege or perishment, most women who study science are handed bad cards in the fight against systemic oppression. The Indian scientific community is losing out on a large proportion of capable scientific minds— the women gold medallists in colleges that we all love to celebrate, those pulled back by socially imposed norms, those who are not wired for rebelliousness and those who have not been 'lucky that way'.

8

Science Spouses

Rebelling against the traditional practice of early marriage is one tactic young women in science employ to stay in the reckoning. If and when they do marry, it becomes important that their spouse is supportive of this tumultuous career. So, what better way than pairing up with someone on the same path? Jonaki Sen is one of the many women in science in India who are married to fellow scientists.

Jonaki left India for New York in 1994 to begin her doctoral studies at Albert Einstein College of Medicine. A year later, she was joined by Amitabha Bandyopadhyay. Fuelled by a common background and passion for the sciences, the two hit it off. 'We just fit very well together,' reminisced Jonaki, 'even when we moved in together, we found that our belongings were complementary! He had a microwave, which I didn't. I had furniture that he didn't have. So it was very nice.' Picking up on the signs, the two decided to marry while they were both PhD students.

Anyone who works in the science community will confirm that scientist couples are commonplace. In conversations we had with women scientists about what enabled their success, we encountered a lot of gratitude for the supportive spouse. This is echoed in a number of surveys that have been published. In one survey published in *Current Science* in 2022, all 115 married women scientists agreed that a supportive husband was critical to managing their personal and professional lives.[1] Clearly, it helps to be in a partnership with someone who can understand the demands and the unusual lifestyle that accompany research. In Jonaki and Amitabha's case, it just so happened that a personal partnership evolved into a professional one.

Having started her degree earlier, Jonaki finished ahead of Amitabha and accepted a postdoctoral position at Harvard Medical School in the lab of Constance Cepko, a leading figure in the developmental biology community. Unbeknownst to Jonaki at the time, this move would come to shape not just her academic career but also Amitabha's. For a year and a half, the couple lived apart, but since Boston was just a three-hour drive away, they could meet often. When it was Amitabha's turn to move to a new lab, his obvious target was Boston. The city being a scientific hotspot, he judged that a suitable position there would be quite achievable. Indeed, he got a bunch of offers, including one from the Cepko lab where Jonaki was working.

But the offer that was most attractive to Amitabha was the one from Clifford Tabin, whose lab was right next door to Connie Cepko's. The Cepko and Tabin labs had been working very closely with each other for many years. Connie worked on the development of the retina and Cliff on limb development, but the former classmates and best friends run their labs as a unit. This is a phenomenon known as 'partner–lab' in the science world. Though they each have their own projects, funds are pooled in, equipment is shared and, often, lab spaces are common. Lab meetings are attended by both teams, so students get a wider understanding than if they were interacting only with their group. Amitabha chose to work in the Tabin lab keeping in mind both his and Janaki's futures. The pair deduced that when the time came to return to India, they would be more employable as a unit if they came with two kinds of expertise rather than just one. For the next few years, though Jonaki and Amitabha had distinct research goals, they worked practically together. The scientific and interpersonal benefits of the Cepko–Tabin model left a deep impression on them.

Today, a similar partner–lab model operates at IIT Kanpur, and it is known to the students of the bioscience department as 'SB Lab' or Sen–Bandyopadhyay lab. Jonaki and Amitabha established the lab in 2006 and, so far, the duo agree that it's been great. 'Our labs share equipment, reagents, bench space, meetings . . . other than the scientific model [Jonaki studies the brain while Amitabha studies the skeletal system]—we share everything,' said Amitabha. He added, 'Our personalities are very different. I am better at lab management; she is better at discipline. Students are more scared of her than me. I don't know why!' he joked, eliciting laughter from Jonaki, who added, 'It's like taking care of a child. I often say I have eight to ten children!'

When we met the duo, they were at the end of a very busy five days, having jointly organized a national meeting that was being hosted by IIT Kanpur that year. Since they work in the same department, they are often called for the same meetings and have the same hours. This is a drawback, they conceded, because their 10-year-old daughter has had to get used to periods of absence. 'She barely saw us these five days!' To their credit, the daughter in question looked very well-adjusted, strolling around cheerfully and confidently mingling with her SB Lab 'siblings'.

A 2010 survey of Indian scientists showed that being married to a fellow scientist is much more common among women scientists than men scientists.[2] While 40 per cent of the women participants had a spouse in science, only 19 per cent of the men did. Having a partner who understands the pressures and demands of a life in science is surely an advantage, but why are there more women in such marriages than men? Over the course of our lab-hopping, we got some clues as to why this might be.

The period after post-graduation and during a PhD is the time most women in science get married. While marriage can spell the end of education for many women, for a lucky few, it is the opposite. This is because for many Indian women, being single does not mean freedom. Most single women in India continue to live with their families much into their adult life, and the patriarchal nature of these families ensures that they are under the control of the father or father figure. Could marriage ever be a way out? Welcome to India's strange 'First get married, then do anything' brand of empowerment.

Condensed matter physicist Vani Vemparala faced this during her early years in science. Fresh out of her MSc at a university near her home, she wanted to pursue doctoral studies abroad. However, being the eldest and most protected daughter, her parents hesitated. To keep her academic goals alive, Vani took up a second master's. After this, she left for the US to pursue the PhD she so wanted. What changed her parents' minds? She laughed: 'I got married. That's what changed.' Vani left the country with her engineer husband who was starting his PhD journey. When we met her, Vani had made a name for herself as a professor of condensed matter physics at the Institute of Mathematical Sciences (IMSc or MatScience) in Chennai.

Similar instances of marriage giving women in science more social mobility are aplenty. For Bhargavi Srinivasulu, an Osmania University-based zoologist, completing a master's degree itself was a big deal. 'I did not have any plans of studying further because I come from a very conservative background. An MSc was it for me; what comes after that, I didn't think about,' said Bhargavi when we met her in the university's nearly 100-year-old zoology department. Marrying her classmate and fellow zoologist Chelmala Srinivasulu was like a shot in the arm for her. 'He said, "I am doing my PhD and I would like you to do your PhD too",' she shared. Inspired by his work on the taxonomy of bats, Bhargavi also immersed herself in the study of these winged mammals. Her family's previous apprehensions disappeared. Eventually, Bhargavi went on to lead her own expeditions into remote areas in search of bats. Her efforts in Kolar, Karnataka, would lead to the rediscovery of a species of bat thought to have been extinct for a decade.

Like Vani and Bhargavi, Seema Pooranchand too was brought up in a protective family set-up. A shy and studious astronomy enthusiast, she was encouraged by her mother to study further and got permission to move into a hostel to pursue her PhD at the prestigious Physical Research Laboratory (PRL) at Ahmedabad. There, Seema took up the challenge of developing an instrument to observe galaxies. She was one of the very few women students on campus. There were no telescopes at PRL at the time, and to observe the Orion Nebula for her research, she would need to travel to various telescopes such as the ones in Japal Rangapur Observatory in Hyderabad and Vainu Bappu Observatory in Kavalur. Her work needed to be done during the night and, sometimes, she faced the horror of being physically harassed by miscreants on two-wheelers who fed off the vulnerability of a woman out alone. However, things improved after Seema befriended, and later married, a fellow grad student. 'He helped me. I had to get attached to somebody, otherwise, I could not go in the night,' she laughed.

While these stories suggest that a science spouse can be empowering for women coming into science, to consider it a magic pill would be a mistake. These positive examples from women in science are probably just the exceptions that prove the rule—the rule that marriage, in most cases, impedes women's progress in science. Especially for the majority who do not end up with science spouses or supportive partners. Despite marriage working out for her career, Vani Vemparala quipped, 'I assure you, no one should get married at 24. People should take their time, irrespective of circumstances.'

The Jonaki–Amitabha and Bhargavi–Srinivasulu examples nicely illustrate how partnering up can foster fruitful scientific partnerships. There are, of course, many more historical examples of this—the most famous being Marie and Pierre Curie who shared a Nobel Prize for their studies on radioactivity. It would seem that for employers, hiring a power couple would be very desirable. Picture this: X is a highly attractive candidate for a faculty position and multiple institutions are competing to employ her. One of them is Institute Q. Institute Q notices that X's spouse Y has an impressive profile too. Institute Q offers both faculty positions, knowing that this would be an offer too good for them to turn down. Now the offer has gotten a lot more desirable for X, and she takes it up because she and Y can be together. Institute Q now has two good hires instead of one, including a woman, which can help improve their gender ratio! Moreover, if a couple is hired together, they are likely to have a deeper commitment to their place of work. The employer Q benefits from 'employee stickiness'. Everybody wins.

Yet, hiring spouses is not a common phenomenon in Indian academia. For a long time, many research institutes have enforced an unwritten rule that bans the hiring of spouses, irrespective of their individual qualifications. Today, the degrees to which institutes are open to the idea vary widely. While some, like IIT Kanpur and many of the IISERs, actively hire couples, in other institutes there are outright (unofficial) bans. We encountered several bitter struggles against the 'two-body problem'—the term that has been given to the predicament science spouses face when trying to secure jobs in the same place.

While hunting for jobs in India in 2004, Jonaki and Amitabha themselves experienced resistance. 'We knew it was going to be difficult,' said Amitabha. 'Firstly, it was a time when Indian academia did not yet have a lot of excellence. It all depended on who you knew and, having been away since 1995, we did not know anyone. We applied to seven institutes. One of the top ones gave us an offer. But they said that they have a policy of not hiring spouses.' This institute informed the duo that only one of them could join; the other would need to look elsewhere. 'We were surprised by the reluctance we faced,' Jonaki said, 'because we were used to the US system where they go out of the way to accommodate spouses if both are good.' After such discouraging experiences, IIT Kanpur's offer came as a breath of fresh air. In Tier-2 or Tier-3 towns where it can be difficult for both spouses to find jobs, it's even more crucial that institutes do everything they can to attract talent. That could be part of the reason why IIT Kanpur was so enthusiastic about hiring Jonaki and Amitabha together. 'There was a red-carpet treatment for us here. They said they had no policy of not hiring spouses, as long as each of us qualified independently.'

Having family members part of the same workplace is a sore topic all over the world because of the potential conflicts of interest that could arise. A report from Stanford University stated that couple hiring was one of the thorniest issues confronting academia in the US, with one department chair commenting that no other aspect of his job arouses as much controversy as hiring spouses.[3] This report showed that the hiring of academic couples had increased from 3 per cent in the 1970s to 13 per cent in the 2000s. It concluded: 'Couple hiring can help build a more diverse, equitable, and competitive workforce, especially with regard to gender.'

The prejudice against science spouses is presumably the outcome of the implementation of the conflict-of-interest clauses in institutional policies. There is also the fear that hiring spouses will be a threat to 'open competition' and hinder the department from making the best possible hire. Amitabha spelt out a scenario: 'Imagine if one of the spouses employed is a rockstar scientist, but the other is not performing. In an attempt to retain the superstar, the institute may end up having to tolerate the underperforming one.'

We also learnt that when a paper is published with both spouses as co-authors, it is sometimes followed by whispers and doubts about which one of the couple is the main contributor and if the other one is just getting 'free credit'. We heard of scenarios where the husband is given credit for a paper he published along with his wife, who was in reality, an equal collaborator. The most famous example of this is the case of Marie Curie. Her husband Pierre and Henri Becquerel were the only people named in the original nomination for the Physics Nobel Prize. If not for one of the committee members protesting this, she would have missed her 1903 Nobel Prize.

A more pressing concern of hiring spouses is that of the welfare of researchers, especially grad students, working under a scientist couple. Devang Mehta, a postdoctoral fellow at the University of Alberta, recalled being advised by a mentor against taking up a project in the lab of a husband–wife pair while studying in India.

'When you're a PhD student, your supervisor has a lot of control over your life. Disputes with them about where to publish your research or how to defend your thesis are very common. It is difficult as it is, but if you have two supervisors who are related to each other, it can get twice as hard. Voting on major decisions having to do with the department can also get biased.'

But Devang believes that an outright ban is draconian. Theoretical physicist Srubabati Goswami is one of many we met who struggled with the two-body problem but made it through to tell the tale. She reminds us: 'After all, any two or more people can team up, right? For example, I can team up at any time with my colleague. Nothing stops people from teaming up anyway.' According to one scientist, banning anything that can be misused is a flawed tendency of the Indian system. 'A few people will do idiotic things. Instead of taking care of this 5 per cent aberrant population, they will bring in rules that hurt the other 95 per cent.'

In 2018, the Bengaluru-based IASc tried to promote a change in mindset by including this line in their list of ethical guidelines: 'There should also be no bias against hiring spouses in the same institution. The Academy should aim for the full and equal participation of women in all academic activities.'[4] Since then, many reports and policy documents have reiterated the need to eliminate the bias against science spouses.

Though some institutions are softening their stance on couple hiring, once hired, couples still have to be cautious to avoid complications. At IMSc, where Vani Vemparala works, we also met particle physicist Indumathi D. and her computer scientist husband R. Ramanujam. Indu and Ramanujam keep their interactions to a minimum while at work and can afford to do so because their fields are distinct. Indu is clear that she would not have considered working in the same institute if they were in the same field. This way, fewer conflicts of interest will come their way. When Ramanujam's name came up in discussions on the directorship of the institute, the couple decided to opt out, thereby avoiding possible complications. 'Say, I got a promotion while he was Director. Others would wonder . . . I would wonder!' Indu said, during an interview at her home in Chennai.

In Jonaki and Amitabha's case, this degree of separation is not possible since their labs work cohesively. The duo have their way of emphasizing that they are not a single unit.

'When one of our PhD students has to face a panel of departmental faculty, we make sure the other one of us is never on that committee. In faculty meetings, we make it very clear that we are two independent members. If there is too much overlap in every sphere—papers, grants—then people won't be able to figure out who has done what. So we try to keep it separate.'

It's not enough that institutes remove their bans on hiring spouses, they also need to supplement it with an inclusive atmosphere. Both in terms of infrastructure, as well as mindsets. Spouse-scientists may require infrastructural support and, childcare, for example. With similar work demands and timings, the question of who will take care of young children is a big one. However, in a situation where a majority of research institutions in the country lack childcare facilities, there are no prizes for guessing whose career is the first casualty between the mother and the father.

Apart from making sure the physical infrastructure is in place, institutes also need to establish an atmosphere of trust. Industrial chemist B. K. Sarojini experienced first-hand what happens when conditions are hostile. Sarojini met her husband at Mangalore University when she was a master's student and he was a research scholar. After completing her degree, she joined the university as a lab assistant. But she realized that having a husband in a senior position was going to be a significant hurdle. She said, 'You have to know one thing. If a husband and wife work in the same institution in the Indian set-up, people will make connections and one of them will have a disadvantage. Here, my husband was a star and I was a lab assistant.'

Sarojini fought to enrol for a PhD as there was no provision for non-teaching staff to pursue research. Moreover, in the absence of a work atmosphere that was transparent, sensitized and encouraging of dialogue, suspicions and feelings of resentment festered. When she finally convinced her guide to accept her as a PhD student, she felt that other teachers were jealous of her motives. They accused her of stealing chemicals from the laboratory, she recalls with retrospective amusement.

'Women are not equal to men, even today. If you publish good work, they think that you have stolen it from somewhere. We have to work 48 hours a day to prove ourselves and I did this.'

Drained by consistent professional discouragement, Sarojini decided to let go of the coveted university job she had held for 13 years to join a new private engineering college as a lecturer. She promised herself that she would return to the university, and when she did, it would be as Professor. In 2013, she did just that. A success story indeed, but the fact remains that it took a long stint at a much smaller college in the city for this science spouse to finally hit her stride. 'There, I started doing crystal studies, getting grants, setting up laboratories. My leadership qualities, my energy came back. I finally became myself.'

The patriarchal institution of marriage is considered an impediment to a woman's career, and this is justified. But it surprised us to see another side of it during our lab-hopping travels. For young women in India who choose their partners, marriage is sometimes seen as a chance for an escape from oppressive family structures. Especially if they can find a compatible partner who is also working in science. Sadly, science spouses are being unfairly penalized by unofficial bans and prejudices that exist within institutional structures. Such bans indiscriminately affect the futures of women scientists, as they are the ones expected to settle for an easier-to-come-by teaching job, or worse, drop out of science altogether, for the sake of keeping the family together.

9

A Hush-Hush Culture

The novel *The Tenth Muse* by Catherine Chung features a mathematician on a quest to solve one of the most famous problems in her field, the Riemann Hypothesis.[1] At one point, early in the protagonist's academic career, we find out about her mixed feelings after winning a prestigious award that recognized her as 'a young woman of extraordinary promise': 'Though I was grateful for the award and the stipend that came with it, all the attention that came with the prize seemed to highlight the fact of my womanhood rather than my talent as if I couldn't hold my own with the rest of the students, who of course were all men.'

Drawing attention to one's 'womanhood' is something that many women in science are wary of. Whenever we sensed a hesitation to speak about gendered experiences, we were reminded of meteorologist Anna Mani's famous comment to sociologist Abha Sur, 'What is this hoopla over women in science?' Yet, as journalists covering gender bias in science, we made it a point to bring this up with our interviewees. We asked them questions such as 'What's it like being a scientist in such a male-dominated field?' or 'What role does gender play in this laboratory?' Many times, we heard responses along the lines of 'science has no gender' and 'I'm not bothered about gender, only about quality work.' This hesitation is something worth exploring, we thought.

In terms of numbers and visibility, women in science stick out. They are aware that their experiences at work can be different from that of cis men, but that doesn't mean that they want to dwell on it—not if they don't have to. As women in science and journalism ourselves, we know as well as anyone that there are far too many bosses and colleagues who are prone to doubting the abilities of those in their midst who don't look or seem like them. In the fight for opportunities, emphasizing that we are no different becomes important for women and other underrepresented groups. This self-preservation strategy works in sync with the prevailing culture in science, where any engagement with social and political issues that affect it is discouraged. As a result, many marginalized scientists have to keep their issues under wraps.

This 'hush-hush' culture allows for the spread of the myth that scientists are immune to social prejudices and that 'science has no gender/caste/prejudice'. We've seen plenty of evidence that those who attempt to bust the myth are penalized. It's no wonder then that scientists are fearful of speaking about the biases they face. The hush-hush culture, thereby, has become a powerful mode of gatekeeping in the science community of India.

We got the first taste of Indian science's toxic hush-hush culture fairly early on in our lab-hopping days. A scientist we interviewed at a government research institute in Kerala got cold feet when the interview was published. She phoned us just minutes after we sent her a link to the report, frenziedly asking for certain parts of the interview to be pulled down. She was worried that she had revealed too much about her work and her musings about inequalities in the workspace. These were the offending quotes: In research outings, I was actually more accepted than men. People in the field gave me a lot more importance and help, seeing I was a woman. Inside offices, however, I see that women need to prove themselves by being doubly efficient. Still, most top positions are occupied by men. Men sometimes find it difficult to accept women—perhaps because they didn't have to interact with women in college. But I think, by interacting with them, we can change them . . . or maybe we shouldn't take it that way . . . if we don't think so much [about not being accepted], approach them again and not take things to heart, it will be better . . .'

Seeing the panic she was in, we pulled down the story. The unpleasant incident illustrated for us how young women in science are groomed to walk on eggshells under authoritarian and opaque management. This scientist's fear of offending colleagues was real and, by no means, unique. There is a reason most scientists quoted in this chapter want to stay anonymous.

'Gender bias is worse in India, and not even subtle, not with some dinosaurs controlling the top positions. We have to be soft enough to qualify as a "good woman", but hard enough to fight for our space,' said Puja*, a mid-career scientist who had become a friend to us over the years. We regularly exchanged memes, interesting studies and frustrations with everyday sexism. For example, once she told us, 'At a meeting, I was consistently called *beta* [son] by a senior scientist.' Puja minces no words when she refers to academia as an 'incestuous world' replete with 'schmoozing parties dubbed as *networking*'. Despite her usual candour, she asked that we not make her name public when quoting her. 'I am brave, but not suicidal,' she said in a message suffixed with a LOL emoji. 'It's too early to take any risks as I'm yet to break into the system. This would destroy my career.'

A. Mani, a mathematician, logician and artificial intelligence researcher based in Kolkata, has observed that being seen as confrontational is something that women scientists want to avoid in order to please the patriarchs in senior positions. Mani herself has faced the consequences of being ambitious enough to address gender and sexuality issues surrounding her trans and lesbian identity. 'I transitioned quite openly and made it a point to explain basics to colleagues through email and mailing lists. Some were explicitly supportive, others did not say anything, and a few were transphobic,' she recalled. In one instance, a senior professor at a central university stubbornly continued to misgender her in emails. Misgendering, the practice of addressing someone with a pronoun, noun, or adjective that inaccurately represents the person's gender or gender identity, is a cause of severe distress to trans persons, often causing lasting emotional effects. 'When I called him out by email, he did not apologize,' she said, adding, 'He played a key role in organizing workshops and conferences in Hyderabad and nearby places. Prior to my transition, I had even been an invited speaker at one such conference, but I haven't been invited since. A fallout of this incident was that I avoided interacting with him, and he stuck to his patriarchal transphobia and religious bigotry. I do face discrimination in Indian academia, especially in relation to jobs and projects, but it is difficult to call them out because of a lack of transparency.'

Behavioural ecologist Bittu K. finds himself in a precarious position not only because of his transgender identity but also due to his activism. 'In India, where people are sort of polite on the face, they rarely come up to you and say why they are discriminating against you. That never happens. People just treat you the way they do and you are left guessing. I have been left guessing whether what I have faced is because of activism or because of being trans.'

As a faculty member at Ashoka University in Haryana, he is grateful to have found a trans-affirming space. He actively works to make the space more welcoming for students across marginalized gender, caste and class locations.

In the south of the country, we met another scientist, but one whose fight for justice was systematically hushed. Rekha* teaches at an engineering college, and we had reached out to her after noticing that she was involved in several local women-empowerment programmes. When we met her in 2017, the engineer was on a protest to bring the authorities' attention to the state of the women's toilet in her department building. We braved her warnings to go see for ourselves and found that she was not exaggerating: the toilet had no seat or lid, there was no water supply, and it looked like no one had used it for months.

During that interview, Rekha had regaled us with her inspiring life journey and her views on the gender gap and other social inequalities. Rekha's incredible resilience and spunk seemed a recipe for success if there ever was one. However, when we spoke a few years later, her demeanour was a far cry from that of the spirited person we first met. She said, 'Some people are mocking me for coordinating women's empowerment programmes while being an engineering faculty member. Because of their public criticism, I have stopped doing this in the institution, for a certain period, even though I am passionate about sensitizing women on many issues.'

Further entangling things for Rekha is her ongoing tussle for a promotion. With an air of defeat, she informed us that there has still been no change in the toilet situation. 'I am silently bearing it along with the other women. I could not continue to strike as I am doing my PhD and there are other risks.'

The culture of silence rampant in Indian science extends to discussions critiquing caste as well. In 1997, renowned mathematician and one of the rare Dalit faculty members at an IIT, Vasantha Kandasamy, filed a case after being denied a promotion for 11 years despite being fully qualified and seeing the posts go to scientists who were reportedly ineligible. She attributed the repeated snubs by the IIT Madras management to their displeasure at her open support for Dalit students on campus facing frequent harassment. Vasantha had always been vocal about the culture at IIT Madras, which has resulted in its evolution into what is sometimes referred to as a 'caste fortress'. It took nearly 20 years, but she was finally vindicated when the Madras High Court ruled that IIT Madras had committed 'gross irregularity' in the selection process.[2] In an interview with a newspaper in 2016, Vasantha, who had retired by then, expressed relief as well as pain: 'I was beginning to despair that I would die without seeing the light at the end of this long battle for justice. The court has now re-established that justice will prevail, ultimately. But then, I wonder if anything can compensate me for the torture, humiliation, frustration and depression I have undergone through these years in spite of my exemplary qualifications and research work hailed across the mathematics world.'

When it comes to sexual harassment, the hushing is especially emphatic. In September 2017, a massive women's rights protest erupted at BHU. It began when a student was molested on campus by men on motorbikes. Security guards who were in the vicinity, and later her hostel warden, all scolded the girl for roaming outside so late (it was 6 p.m.). The warden reportedly told her: *'Tum ladkiyan itni der tak bahar ghumogi aur kuch* *hota hai to aa jati ho complain leke'* [you girls roam outside so late and then come to us with complaints when something happens]. Further, the authorities reacted by advancing the curfew for women from 7 p.m. to 6 p.m. This gross injustice was too much for this student and her peers at BHU, so they decided to fight back. At several peaceful protests, the police reportedly used force to disperse the protesters. The incident was in the limelight because it occurred during the week of a visit by Prime Minister Narendra Modi to his constituency of Varanasi. Payal*, a PhD student at the university who was part of this movement, wanted to do her bit by reporting the happenings at BHU on our website thelifeofscience.com. But hours after her article was posted, she was advised that this act of hers could put her in harm. We removed her byline.

What we drew from these experiences is that academia is far from an environment that values freedom of expression. And while it may be easy for the more privileged of us to comply with the rules of 'hush-hush', for underrepresented groups, bringing up issues is critical to their very existence in science—it may be as fundamental as having one's identity acknowledged, having access to a functioning toilet or feeling safe on campus. Allowing the system to 'hush' women in the above ways is as good as telling them to shut up and bear the injustices, or else, leave science.

In a scenario where acknowledging gender bias is risky for scientists, it's no wonder that the 'F word', feminism, gets an even worse rap. Even scientists who are outwardly advocates of gender equality hesitate to align themselves with this label. When we posed the 'do you consider yourself a feminist' question to the scientists we interviewed, only 31 of the 67 who responded said Yes. 15 said Maybe and 21 said No. However, a more detailed look at the Nos and Maybes revealed something interesting.

No, you should be assessed by your skills and achievements and not by gender.

No, but I promote females and wish to see many more women in science at higher positions.

No. I believe in recognizing talent irrespective of gender.

Maybe. I believe in social equality of all human beings irrespective of gender, caste, religion, financial background, sexual orientation, education, class and geographical location.

Maybe. If it's about standing for unfairness then I can be one. But if it means fighting for reservations, then I am not.

Maybe. I believe in equality or at least equal division of responsibilities. It does not mean I hate men but I do despise patriarchal mindsets that dumb women down to being livestock.

These Nos and most of the Maybes were evidence that even though the demands for equality from women in science are getting louder and more intersectional, there is a reluctance to label their outlooks as 'feminism'. Feminism is associated with hating men, demanding special treatment, or being an excuse to reject tradition. The lack of education about gender and gender bias in our country has resulted in generations of Indians who have not understood that feminism is the ideology that opposes patriarchy, a societal reality that limits not just women, but also men and other marginalized groups. Feminism, in its true intersectional form, works together with other movements for social equality like the economic, anti-caste, sexuality and disability movements. Being a feminist does not necessarily mean rejecting the roles of mother or wife; it only advocates the rights of women to have other identities instead, or in addition to these societally approved ones. Nobody tells us this, so naturally, there is a great hesitation to accept the label 'feminist'.

Some women scientists feel that it may not be sensible to walk around freely identifying as a feminist. In a deeply patriarchal society where 'feminist' is often used as a slur, they fear it might work against them. 'I feel the tax to be paid is very heavy with this tag,' said a young biologist who did not want to be named. In her career, Rachna* felt an unexpected pressure to conform to some stereotypes. She explained, 'Outspoken women scientists need to compromise by talking about their personal lives. This seems to me the biggest tax to pay if we want everyone to become aware of existing challenges. People only understand when you reveal how hard it was. However, I do not wish for my life to be an open book. Unfortunately, this is a contradiction to what is required.' Despite this, Rachna faced some harsh realities that made her rethink her stance.

'As a PI, I am witnessing huge problems involving women. One of my postdocs was not given an extension when she was on maternity leave. She stayed with me and we published a very high-impact paper together. Another student was beaten at home by her family when she worked late one day. In these scenarios, you cannot hold back your urge to speak. Of late, I have realized that we have to pay the tax of speaking out for under-representation, else the next generation will have to reinvent the wheel . . .'

Are the risks milder for men scientists who are vocal about gender equality? Perhaps, but they are not immune to garnering problematic reputations. Ram Ramaswamy, a physicist and computational biologist, and former President of the IASc, recalled his late spouse Charusita Chakravarty's concerns about his active engagement with women in science initiatives, though she was a scientist too.

'She was on many committees and herself was very vocal, but she would ask me why I wanted to become a gender activist at the expense of science. She used to be concerned that people will take me less seriously as a scientist.'

Ram knew Charusita's fears were well-founded, but he believed that as long as he kept actively doing good science, it wouldn't be a problem. He went on to occupy many leadership roles in science and this power allowed him to form his own identity. 'Many women who speak out are viewed as whiners using this [gender issues] to move ahead. For some of them, the activist label overcame their scientific label.'

Mukund Thattai, a biologist and a popular and progressive voice from the National Centre for Biological Sciences (NCBS) in Bengaluru, affirmed that being seen as rocking the boat can be risky. Yet, he suspects that the silencing may not be as bad as it is made out to be. 'There are no mechanisms in place that are muffling people. There is an impression that it is much worse than it actually is, though it's still worse than it ideally should be,' he said. 'Maybe you won't get called for the next meeting!' He recalled at least one instance where his speaking out has had a positive outcome: 'Last year, when I was invited to a meeting with no women, I wrote to the organizers about it and they took it seriously.' But at other times, he has had to give in to the silencing.

'I was once on a committee with only one woman. The chair of this committee (a man) would comment on what she was wearing. At that point, I wrote to a senior and the woman concerned asking if we should do something about this. But I was advised to let it go, as the chair was an old person, from a different generation. Later, I realized that there was nothing to accomplish there [in the case of the sexist committee chair]. Even if there was an apology, what could be done with that?'

Whether or not you can get away with bringing up issues, according to Mukund, depends on many factors, an important one being the stage of your career.

'If after putting yourself out there, you don't see support, then you are less likely to bring it up the next time. Junior faculty may not have this support. There needs to be a critical mass of people willing to support you.'

Would he advise a junior to speak out against any sexism she faces, say, during an interview? 'This has actually happened,' he replied immediately, and added, 'And my advice to her was to hold back. I can't sit on my high horse and advise someone to take a risk in a context that I don't know of. It doesn't help.'

Being an advocate for gender equality is not something scientists are comfortable flaunting, and this isn't always conditional on how professionally established they are. On the one hand, younger scientists gearing up to attain relevance in the competitive science world fear that opening up about discrimination or difficulties will brand them weak; on the other hand, senior women scientists fear that such comments will elicit doubts that they rose to the top by 'playing politics' or asking for favours. Speaking about gender could get senior scientists distanced from the old boys' clubs that they have worked hard to earn a place in. Being kicked out of the club means that they will lose not just their status, but also the potential to be an agent for change from within.

Renowned biologist Sandhya Visweswariah from IISc shared her perspective on the feminist 'tax' during an interview at an international cell biology meeting. Despite being a strong advocate for women-friendly policies, she admitted this when asked what senior-level women like herself could do to close the gender gap: 'Not much, actually, and the reason is, if you start making a big noise, you are considered a feminist and then you become strident and they avoid you. They don't want to talk to you after that, right? It's very difficult.'

Another scientist, who did not want to be named, confirmed the risks but added that this doesn't stop her from openly identifying as a feminist. 'It's true . . . in fact, not only do they avoid you, they undermine you as a scholar and a member of the community. But the responsible reaction to finding things wrong with the community is to work towards setting them right. This is even truer in a publicly funded institution. If I am paid by taxpayers' funds, I owe it to the taxpayers to not only deliver on my narrow mandate of rigorously researching, say, organic chemistry, but also on the broader mandate of empowering the next generation and institution-building. I may suffer in my personal career growth but I would have left behind a more meaningful legacy.'

When women on top do take bold stands against the hush-hush culture, the impact can be momentous. We witnessed such an instance in Kolkata at a programme for mid-career scientists selected for their leadership potential. The audience consisted mainly of male scientists. Senior particle physicist and Padma Shri awardee Rohini Godbole had been invited to speak about the gender gap in Indian science. After her talk, one of the participants brought up the rhetoric of 'affirmative action will dilute excellence'. For long-time gender equity advocate Rohini, this struck a raw nerve. She retorted fiercely, daring him to voice what he was implying—that women scientists are less 'excellent'. The questioner saw his folly and sat back down, embarrassed. It was probably the very first time these men were being confronted for holding sexist mindsets. The few women scientists in the room that day left with their heads held a little higher than usual.

10

Reframing the Leaky Pipeline

The metaphor of the 'leaky pipeline' has been used extensively to explain the gender gap in science. 'The pipeline is leaky', 'plug the pipeline or we'll lose the women,' it is often said. This choice of words doesn't fall easily on the ears of many Indian women in science. Another description inspiring snubs from many women we met is 'dropouts'. Being referred to as a potential 'drip' or being seen as a 'leak' that falls into the gutter to be wasted away is not something one is inclined towards. On the contrary, when women leave the lab for good, they are often making empowered choices and looking out for themselves and their kin. They opt for a different setting, away from the academic track, freeing themselves from the visions of becoming a scientist that they once held.

The bitter truth is that a life in Indian science can be hard to find for a woman and once it is found, it is hard to sustain, demanding a lot more from women than from men. Greener pastures for women who leave the institutional track may mean a better life, a better job or as we found in some cases . . . better science! Dreams and hopes are indeed betrayed when women leave academic science and research, but more severely, it is our scientific ecosystem that suffers their absence. Science, while intending to be an open, diverse and self-correcting enterprise, continues to operate as a monolith composed of upper-caste cis men, their research questions and findings that very sparsely find translations into society.

One consistent observation in wide-ranging studies of organizational behaviour conducted in occupations such as homemaking, administration, factories and farm collectives, even in boardrooms, is that women tend to collaborate more than men. Not only are women more inclined to collaborate, but they also carry a collaborative overload. Collaboration overload is a term used by researchers studying collaborative work cultures, more recently in the US, to explain the decrease in performance of an individual due to their high involvement in meetings, addressing requests for inputs and advice and setting up helpful structures on behalf of their organizations. One such study published in the *Harvard Business Review* in 2018 reports: 'We recently studied three such organizations in-depth using a combination of surveys, interviews, and direct observation, and we noted a consistent theme: While everyone in the organization experienced collaboration overload, women felt the burden disproportionately.'[1]

Even back home, studies suggest that women in science are overloaded. Namrata Gupta and A. K. Sharma are sociologists who write in their paper 'Triple Burden on Women Academic Scientists' about the multiple roles that women in science tend to have. Identifying oneself only as a scientist is unaffordable, especially for women fulfilling traditional roles in society, of a mother, a daughter-in-law and a wife. 'The multiple roles women have to play lead to efforts at time management and self-discipline that leaves them with the feeling that life has become mechanical,' Namrata and

A. K. write.[2] In a survey they conducted, the duo asked the women they were interviewing to score their tiredness. Their results read: 'Average score of physical tiredness was 3.32 out of 5 and all kinds of exhaustion covered—emotional and mental—were also above the halfway point.'[3]

So, do women leave science because they tend to be overloaded and tired?

It seems ironic that women, a group shown to be more collaborative, are missing from science, which is a domain that today thrives on collaboration. Gone are the days of the lone scientific genius in his ivory tower. The chances we will ever hear of a naked man scientist running amok on the streets shouting 'Eureka!' again are very slim. Science breakthroughs are hardly ever due to an individual feat. We are in the age of 'Big Science'; our efforts to disentangle questions at the edge of science are global collaborations involving hundreds of scientific minds, often mediated through geopolitical diplomacy. An example of mega-global scientific collaborations is found at the International Thermonuclear Experimental Reactor (ITER), where the promise of a clean and secure global energy in nuclear fusion (the same reaction that energizes the sun) is being sustained and is also seen in the successes of the International Space Station and the Human Genome Project. When the discovery of the Higgs Boson—sometimes dubbed the 'God particle'—won two (male) physicists the Nobel Prize in 2013, there was a lot of discontent at the committee's overlooking of the fact that it was the giant effort of thousands of scientists at CERN that resulted in the breakthrough.[4] The necessity of collaboration in science was further emphasized during the pandemic when it became crucial that countries share data from clinical trials, as well as in the pooling of genetic sequences of the coronavirus variants.

Breakthrough scientific results that expand our understanding of the world in a significant way are always achieved as a group. However, this reality does not seem to have dawned on the science culture inside the labs in our backyards. Our science institutes still uphold the idea of the lone science slog. An image unpalatable to most women who, more often than not, take on multiple roles in their lives.

Do women then leave science because its culture is not as collaborative as the work intends it to be?

To find the right answers to the question 'Why do women leave science?', data on those who leave science is required. This is largely missing. It was completely missing before Anitha Kurup, a sociologist at the National Institute of Advanced Studies (NIAS), Bengaluru, got to work on the question. 'I decided to do this because I'm a feminist,' she told us on a boat cruise on the Brahmaputra in Assam, a recreational element of a scientific meeting. 'It was extremely hard to trace the women who have doctorates in science but are not currently working. I spent two-thirds of the study time convincing dropouts to participate in the survey,' she said as we munched fried snacks with ketchup on the upper deck.

The research Anitha was referring to is named 'Trained Scientific Women Power: How Much Are We Losing and Why?' It was published by the IASc in 2010.[5] Anitha, a sociologist in the research team, led the study that built a database of around 2000 women and 400 men possessing science doctorates, and then studied their different circumstances through surveys. The study is unique in two aspects. First, it acknowledges that gendered experiences of women in science are varied and the solutions to close the gender gap must also work multifariously. The study segregates the responses from women with science PhDs into three categories—women in research, women not in research and women not working. Secondly, the survey also made room for responses by men in research. 'If you don't talk to the men, you don't know the rules,' Anitha whispered as we floated further down the river.

The report has many interesting findings. For example, it finds that 14 per cent of active women scientists are single, compared to only 2 per cent of men scientists. 86 per cent of men had children compared to the lesser 74 per cent of women. Perhaps the extra hours of work women have to put in repels women with kids and married women from science, as 46.8 per cent of women compared to 33.5 per cent of men reported working between 40 and 60 hours a week. These results point to the reality that science research institutions in the country favour women who come with no strings attached. Men's participation in science—whether married, single or childless—is relatively unchallenged. Compared to women (as a group), science is smooth sailing for men.

Science, as it is today, is a male occupation. Dominated by men, scientific institutes are neither suited for nor proactive about accommodating any gender identities except cis men. The crudest example of this is the lack of ladies' toilets in most scientific institutes and their deplorable quality even when they are present. While toilets have been rightly recognized as a major factor in adolescent girls leaving school in India, not as much attention has been paid to women's toilets in workplaces.[6] Several times, department buildings in universities do not have easily accessible or sufficient ladies' toilets. As we have seen in one example from a previous chapter, often the toilets that do exist are thoughtlessly designed, without conveniences such as a water supply, a hook to hang one's bag, a surface to place sanitary items or a dustbin to dispose of used sanitary products. Feeding stations for young parents that work in our labs are also missing. From our visits, it was clear to us that infrastructure committees inside the institutes we were visiting had scant representation from women, disabled and trans communities. There seems to be a lag between claims of gender equality and the basic infrastructure our institutes offer.

Jahnavi Punekar, a palaeontology faculty at IIT Bombay, confirmed the situation from her student days at her institute.

'I have lived in both the female hostels at IIT Bombay when I was a student here a decade ago. The facilities including toilets in the older hostels were TERRIBLE. The new one was fairly good.'

When Jahnavi joined the Earth Sciences Department at IIT Bombay in 2017 as faculty, one of her first responsibilities was to press the administration to provide better toilet facilities for women in the department:

'I have been the only female faculty for a while and more or less the first one in the Earth Sciences department at IIT Bombay. When I joined, the female staff and students did come to me complaining about the mess inside the toilets and the lack of soap, dustbins and tissues.'

As lab-hoppers, sometimes, we were ourselves inconvenienced. In December 2018, when we spent two days working at the media centre at the IIT Kanpur campus, a polite sign in the ladies' toilet asked us for too much. It said: Please avoid 'shitting', as the flush isn't powerful enough. Another time, a physics and gender conference boasted of 60 per cent of the delegates identifying as women. The meeting hosted by the University of Hyderabad called itself 'Pressing for Progress' while pressing to relieve oneself at the conference venue proved tricky since the toilets lacked water supply. Some delegates chose to use the men's toilets, as they were better equipped.

The Right to Education Act of India 2009 and National Urban Sanitation Policy prescribe a toilet-to-user ratio of 1 toilet per 40 users. The reality of our ladies' toilets at state-level universities and colleges is appalling. A 2016 survey of 10 women's colleges in Jaipur reports a toilet ratio of 1:232 for students and 1:47 for lecturers. In the same survey of 270 respondents including lecturers, students, housekeeping supervisors and helpers, 96.5 per cent said there were taps in toilets but only 30 per cent of lecturers and 29 per cent of students said that there was water in these taps.[7]

Women leaving scientific work is a result of a deficit that lies not in the women themselves but the basic infrastructure at our institutions. This deficit starts with, not in the least, toilets. Along the way, we met women on the brink of leaving science because of other infrastructural scarcities.

One such woman at the crossroads was Vineeta.* At the time of her interview, she was a postdoctoral fellow in Hyderabad, where she worked on several research projects in a biology lab. Vineeta started to tell us about her conundrum by sharing a very difficult conversation she once had with her husband.

'We were in the car. I asked him, "I have so much training in research. How can I leave behind everything I learnt before?" And then he said to me: "It is up to you. In my life, I have realized that my wife loves research more than she loves me." That made me feel so bad,' she said whilst laughing and crying simultaneously. 'I know that he wasn't saying I don't love him but that he missed me.'

For the last 10 years, Vineeta had lived away from Arun, her husband of 15 years, a Colonel in the Indian army. This arrangement that had proved to be intolerable to the young family transpired because of Vineeta's love for research, she said. As army persons often are, Arun is posted at a different cantonment every two years. Vineeta, although dazzled by the life of an army man's wife, could not stay away from her experiments. In all the years of separation, Vineeta was up to standardizing a scientific protocol to work on exosomes—bubble-like cell organs containing a telltale cargo that is a goldmine for cell biologists like Vineeta.

We met Vineeta for the first time at a workshop on 'science journalism for women' in Bengaluru. She popped up with an earnest question: 'You have interviewed many women in science, tell me, do others also worry about their child's mental health?' Stumped by the question, but intrigued, we went back to her later to hear her story.

Vineeta and Arun's son Amey was eight years old at the time of the interview. All these years, the family has not lived together because the professional lives of the parents did not allow it. The negotiation between Vineeta the scientist and Arun the Colonel, on how to keep their family together has dragged on longer than any of them have the forbearance for. To Vineeta, it seemed, they had reached a stalemate. If both continued with their jobs, the family would stay separated for too long, which, according to Vineeta, could be a threat to Amey's overall development in his crucial growing years. 'I want Amey to have both his mother and father around, so he can have a balanced view of his own life,' she believes.

'Amey is at the stage where he is learning to mimic adults. His mirror neurons are firing. So as a mother if I don't have enough time and neither does his father, then we are being unfair to our son.

I want my family to be happy and together. I can't forgive myself for leaving my husband to live all alone for so many years, and not giving him the privilege of being with his son. We don't want to do that anymore.'

So, Vineeta had been scouting other career options like science communication, in hopes of a future shift to a job with more flexibility. Wrapping up her career in scientific research to rejoin her family was as imminent to Vineeta as it was devastating. It didn't help the science side of things either that an army life would come with perks that Vineeta felt the family deserved after all these years of separation. 'There are many conveniences like *dhobiwala* (laundry people), *safaiwala* (cleaners) and cooks if we live with Arun. Here in Hyderabad, I am alone with my son without any support,' she said.

'My husband has given me an ultimatum. I have to make a choice very soon—science or my sanity.' As a postdoc over the age limit for entry-level faculty positions, Vineeta felt her life in science had become sort of a gamble. At her institute, she relied on yearly extensions to her postdoc position.

'It's not easy. I need to run behind my career, which is exhausting. And I won't even get a permanent position because I started my PhD late and hence, I am past the age limits for many of the entry-level jobs. Taking care of my son all alone is really hard—the quality of daycare at the institute does not suffice.'

'I don't want to call myself a dropout but I see no choice in front of me but to move away from scientific research.' The lack of basic support for good quality childcare, family care and age-flexible policies in science, sadly ended up failing Vineeta the cell biologist.

The reason Vineeta eventually decided to quit the lab is the lack of support from scientific institutions for people who choose to be primary caregivers. 'If married women become scientists, there is a presumption that they have ample support at home.' The result of this bias is that the research institute finds no need to instate infrastructure for childcare, elder care or family care. It is, therefore, no surprise that in our reporting of success stories of Indian women in science, a vast majority of women benefited from the support of parents, wealth or lineage—upper caste or upper class. These privileges are entry passes in the gatekeeping underway that keeps many marginalized women out of science. In such a scenario, where most of our brilliant women scientists are either from such privileged backgrounds or can manifest a super-heroine strength to rebel against society, women like Vineeta take all the pressure.

'Women with kids should be considered a very normal demographic. The lack of family-friendly policies is linked to the "privileged married woman" stereotype. Many of us prospective "dropouts" suffer from it. Why can't we come to work? For me, that is the place where gender bias happens,' Vineeta protested.

In Anitha's survey study too, the lack of institutional support showed up as one reason for women leaving science. One of the conclusions of the 2010 study says:

> A significantly higher percentage of women-not-working reported having received no help with childcare [while they were working]. Thus, for women not working, the absence of support either by choice or compulsion could perhaps be an important reason for their dropping out of science.

Vineeta has put forward the case for institutional support for women with families at several postdoctoral meetings and also in an opinion piece she has written. She has campaigned that small considerations would make a world of difference for women like her and, in fact, everyone with families working at research institutes. Some of these are health insurance coverage that extends to families, well-maintained high-quality daycare, programmes for women postdocs and training to make up for lost time fulfilling family responsibilities. When her opinions were published, they immediately received retorts from established women in science. They argued that 'freebies' for women in science were a step in the wrong direction and that no amount of daycare centres and special schemes would help women in science. 'We need equal opportunity. Period,' they said.

The incident revealed to us the fractures within the women-in-science movement in India. On reflection, we realized that stopping at the demand for equal opportunities for all will not help women who need institutional support now. In India, many patriarchies are at play in the social fabric—some in relationships with parents, several with partners and, furthermore, in academic institutions. We require multiple feminisms to challenge them. Expecting all women to fight on all fronts, skirts misogyny by turning a blind eye to ground realities.

While the paradigms of Indian masculinity change over long periods and men are expected to take up the unpaid caregiving roles that women have historically borne, it would be futile to simply wait and do nothing as women scientists with family responsibilities are forced to leave science. Our institutions can immediately help them by taking affirmative action, such as relaxing age limits on tenures or grant applications, transparent hiring procedures and keeping a check on biases at play and providing quality daycare and toilets; in short, be prepared with whatever they might need to stay on and thrive. These measures have to be implemented in a way that is inclusive of all groups that need them.

Keeping in touch with the scientists we met meant that we got to witness their careers evolving over the years. Some won awards, some finally got that permanent job they were coveting, some had children and took a break, and so on. When we checked in with Vineeta at the end of 2021, she had moved to Bengaluru, was happy for her family and is still job-hunting. In a handful of cases, the women left academic research—not for their families but, ironically, for science. One of them is Soumya Prasad, a conservation ecologist who did her PhD in the scientifically fertile ground of IISc in her home city of Bengaluru. She was then a faculty member at Jawaharlal Nehru University (JNU) in Delhi but in a few years, gave it all up to live and work in Dehradun. She now runs a sustainability centre aimed at communicating the grave messages of ecological collapse and is driving her community towards scientifically founded local solutions to counter it.

'Ecologists are always feminists,' says Soumya. She is one of the many ecofeminists in the country who opted out of the prescribed academic pipeline to pursue applied work with ecology.

'Indian institutional science in many ways keeps women out but . . . it also is . . . I don't know the word to use . . . "anti-feminine". Its systems are not suited to how women usually work. Women generally like to collaborate. Even in their homes . . . wherever you see their work in society they come together to do things. And in my experience, it is very difficult to collaborate at Indian academic institutes.'

When we met Soumya for the first time at JNU in 2016, she had freshly bagged the Assistant Professor position thanks to a faculty recharge programme, instituted by the UGC to infuse fresh blood into the central university system. Her 'lab' consisted of a few rooms with cubicles for her students in the building meant for the Special Centre of Nano Sciences, not an ecologist's natural habitat. Back then, she had spoken to us excitedly about the discoveries she made in the Rajaji National Park on seed dispersal adaptation by trees in response to climate change. She seemed pumped up for a productive future at JNU.

Two years later, we met her in Dehradun at the sustainability centre named Nature Science Initiative she founded with her husband. The JNU job was now behind her. As always, she was enthusiastic and spirited about her work, but this time Soumya's musings were flavoured with the mature humility of someone who had prevailed through difficult odds. When she talked about Indian academia, her words bore a sense of frustration and an attitude of 'good riddance'. What had happened?

'The more I think about the life that I was leading at JNU . . . the more futile it seems.'

'Initially, I used to turn up at the office at 8 a.m., which is when I typically start working. I believe in starting the day afresh, with new energy. But when I got to the office, nobody else was around since I was not within the Life Sciences Building.'

To make matters worse, the office premises that were offered to her provided no Internet and no phone connectivity.

'It was weird because as an ecologist with projects underway in different parts of the country, I need to stay in touch with my field teams and students on and off the field.' Soumya said. The daily anticlimactic rhythm of entering the office to the upbeat sound of new possibilities quickly flattening out on finding no connectivity sucked something out of her.

Soumya grew very tired, very isolated and very bored. 'It was so depressing.'

When Soumya started to spend more time at her quarters, things began to be bearable again. At least work would go forward. 'I could work from home, thanks to an internet connection I had privately arranged for myself. I just went to the office to sign papers,' she said.

'I love doing my work, unfortunately,' she said sarcastically. 'I just love my work! I can't stop doing it. And in this Indian academic context, I was unable to do it.' Soumya longed to get ahead in her discipline of applied ecology. This work, she found, was increasingly hard to do in the university set-up. Her research group disbanded eventually at a loss to the university and Indian ecology. 'I think there's a serious problem with the way we do higher education in the country. There's a very very serious problem with it.' Soumya's research interest in applied ecology is the need of the hour for the country. Indian cities especially have gone through multiple ecological collapses in recent years; Bengaluru's disappearing lakes and Delhi's grey winters are testaments to that. Be it addressing issues in waste management or preserving forest ecosystems in the face of climate change—the urgent need for applied ecology work in India cannot be overstated. Researchers like Soumya offer solutions to address environmental problems at larger scales with very limited cost and time investment. For example, one of Soumya's initiatives works towards raising awareness about the risks of plastic ingestion to endangered animals such as elephants and the greater adjutant stork, which have become dependent on feeding at open landfills.

It is important to consider this: Soumya's work continues, although differently. 'Applied ecology remains today in the domains and realms of a handful of NGOs. There's no serious engagement by any of the government scientific institutions,' Soumya said emphatically. Her anger was palpable when she went on to tell us about the time when a funding agency turned down one of Soumya's research projects on grounds that they 'don't have money for this kind of NGO work'.

'There's a problem with IIMs, IISc, IITs . . . all of these. They rarely connect with issues, or the societies or the environment that they live in. Why . . . why don't they do this? Only then would they be of any value to society or the environment. There is a new IIM in Kashipur in Uttarakhand. I tried to engage with them about community-based tourism, which is part of the really important commitment of Uttarakhand to ensure that local people benefit from tourism. But IIM Kashipur doesn't engage with sustainable tourism in Uttarakhand. And even IIM Bangalore, for example, doesn't want to engage with traffic issues in Bengaluru. This needs to change.'

Soumya counts herself among those who have no place within the infrastructure of our academic institutions, in its present form. Taking herself and her important ecological work 'out of the system', so to speak, became critical for her to manage to do the work. Soumya and many other women like her consciously move into a sphere of science-based social engagement that institutions, for no good reason, ignore.

Our encounters with women in science who have left institutional science reflect that they often make the choice to work outside these institutions. 'Indian research institutes don't support women because institutes are not collaborative and the work being done is not translatable into our society, environment and communities,' Soumya summed up.

One could argue that the scarcity of connectivity and collaboration that Soumya faced is a gender-neutral experience within the academic structure. However, we maintain that connectivity and collaboration are part of the basic institutional support needed for true inclusivity. Just as Vineeta's life in the lab succumbed to the lack of institutional support, Soumya's life in academia came to the same end. Both their choices to formally leave institutions were inspired by their gendered realities. Most upper-caste men, however, face much fewer hurdles. Their career trajectories, on the other hand, are fast-tracked into higher academic positions while being buffered by a much more forgiving support system, leaving the women to trace their unique paths.

Soumya's example also serves to problematize 'the leaky pipeline' model to explain the exodus of women from institutional science. A global team of STEM researchers write in their contemporary (2021) retort to the leaky pipeline model:[8]

> The traditional 'pipeline' model of workforce development no longer captures the reality of a modern STEM career. Nor does it represent the broad range of interdisciplinary and innovative opportunities that STEM professions now offer across a wide range of industries, nonprofits and other organizations. The pipeline model has also contributed to the historical exclusion of individuals from minoritized ethnic, racial, sexual and gender identity, disability and service status communities.

An alternative is suggested, next.

> It is time for us to dismantle the pipeline. Instead, let's take inspiration from the natural world we study and envision a new model that captures the opportunity, variability, and responsiveness of a modern STEM career; that embraces the diversity and experiences of the people who engage in it; that recognizes the many on-ramps, pathways, and career pivots that real life induces; and that provides a framework in which there is a place in STEM for everyone. This is the braided river model.

A nod to the braided river metaphor was seen in a publication published in 2022 by UNESCO India, titled *The Braided River:*

The Universe of Indian Women in Science, in which several science women are seen in their element.[9]

However, framing it differently can only go so far. In the attempt to evolve the narrative around women in science, we must be careful that we don't end up normalizing the fact that women scientists struggle to find space inside our scientific institutions. The gender gap persists regardless of the choice of metaphor. In the quest for bridging the gender gap in Indian science, we cannot go on without addressing the deeply rooted biases that are behind it. We take these on in the next part.

Part III

Noticing Patterns

11

Prejudiced Remarks and Silently Sexist Thoughts

As any woman scientist will tell you, there is a widespread societal 'mindset' dictating that a woman's place is not in the lab. It is at home. 'A scientific life is not for women.' Even though women have been challenging this stereotype throughout history every day, the 'mindset' will not budge.

The mindset positions a cis man as the most suitable candidate for probing the laws of nature and expanding human capabilities. This fits neatly with the disproportionately large number of 'fathers' of scientific disciplines, such as Gregor Mendel, the Father of Genetics, Galileo, the Father of Astronomy and no less, even the Father of Modern Science, a title contested by none other than our default figure of scientific genius: Albert Einstein. In today's more multidisciplinary fields, we have the likes of John Tyndall, the Father of Climate Change; Hans Selye, the Father of Stress Research and so on . . . In India too, there are no 'mothers' to be found among Sushruta, the Father of Surgery; M. S. Swaminathan, the Father of the Green Revolution and P. C. Mahalanobis, the Father of Indian Statistics. Even in the 2021 book *Founders of Modern Science in India*, authored by famous Indian scientist C. N. Rao and his educator wife Indumati Rao, out of the 16 founders profiled, there is only one woman. And she, a pioneering expert on chromosomes and hybrid varieties of plants, has been decorated in the book with nurturing descriptors such as 'Sugar Queen Janaki Ammal, She Made Our Sugar Sweet'.[1]

The default image of a scientist as a father figure implies that mothers only give birth to humans, while men spawn not just human children but also brave new ideas that change the world. This mindset that we borrowed from history is now cemented in our conditioning by repetition, a disservice continued by popular science media. It is also deeply ingrained in our institutions and omnipresent in our labs. We have been force-fed the gendered idea of a scientist in the image of . . . a lone genius . . . the wise old man who knows best.

Janaki Ammal in the 1950s

Women and nonbinary people working in labs today are at the receiving end of severe marginalization caused by this very mindset. Lab-hopping brought us to numerous situations where we witnessed the dehumanizing treatment of women in Indian science. The mindset often manifested as microaggressions that we were blind to before we deliberately chose to put on feminist glasses and stopped expecting to see wise, old boffins as the heroes of all of humanity's (science) stories. And the stories we wrote whilst lab-hopping met this mindset head-on. At the conferences we attended, the mindset often translated into a bias and jumped out like an annoying jack-in-the-box. At these events, cis men take up all the space and insist on having the last word as scientific authorities, leaving no room for critique or discussion. Below is an example of the bias playing out against women from our report from a scientific conference in Dharamshala:[2]

> People in rural Kullu, are bound to the forests and importantly their protection through their local deities (demigods) along with the rule of law. The fear of God is helping conserve the biodiversity of the national park,' Monika said, concluding her presentation on her most recent work on the influence of local traditions in the conservation of the Great Himalayan National Park.
>
> The end of Monika's presentation signalled the beginning of the Q&A session. It started with the shaking heads of the two senior scientists, both male, chairing the scientific show-and-tell at Him Science Conference. They were both making the same point; they protested that relying on traditional knowledge for any conservation efforts would be a 'step backwards'. A discussion ensued between them, with Monika trying to get a word in unsuccessfully.
>
> The five minutes reserved for Q&A with Monika, an opportunity for her to defend her research, were hijacked as the rest of us looked on, slightly irked.

Along the way, we heard several accounts of how biases play out in the lives of many marginalized identities in science, and how the gatekeeping underway, more often than not, succeeds in keeping diversity out of Indian research. The stories we collected are the realities behind the grim gender gap statistics in India. Whenever we asked Indian women in science what the reason was behind the gender gap, the answer always was 'the mindset' or 'the bias'. Their responses reflect the Indian brand of gendered realities, but we do share its origins with the rest of the world in the history of science. Behind what our interviewees dub as 'the mindset' is a globally prevalent bias against women pursuing knowledge. The bias runs its course along a timeline that is punctuated by doors of universities and thinking societies slamming shut on everyone but gentlemen and sages. It is realized in the lives (and death) of women like Hypatia—a Greek astronomer and mathematician brutally murdered by fanatics on the steps of an Alexandrian library, and in the choices of many women intellectuals including one of Galileo's daughters who became a nun to afford the life of the mind. This bias is also seen in stories and experiences of several women in history including Plato's women students dressing up as men to camouflage their way into education, and the many women like Rosalind Franklin who have had their science stolen by men, and the thousands of women scientists like Jocelyn Bell who never receive the positions and/or awards they rightly deserve. Closer home, we see the bias reflected in the success stories of Rajinder Jeet Hans-Gill who wore a turban and a boy's uniform on her road to becoming a mathematician and in the triumphant Kamala Sohonie whose satyagraha opened the gates of IISc Bangalore for women.[3]

Over the recent decades, this bias has been studied extensively. A small study by three psychologists we interviewed in 2016 analysed the results of different surveys including a dataset of around three lakh respondents that showed Indians harbour both implicit AND explicit biases against women in science.[4] Implicit biases are mental processes that exist largely outside of conscious awareness and control (for example, by default addressing formal emails to 'Sir' assuming a boss is a man—a practice far too common in India[5])—whereas explicit

Students outside IISC's first women's hostel, c. 1945. (L-R) Rajeswari Chatterjee, Roshan Irani, M. Premabai, Miriam George and Violet D'Souza

bias is largely conscious and could be expressed in words, actions and choices (for example, declaring that 'women don't like physics' as a suspended CERN scientist infamously said[6]). Their study also draws a link between the bias and government data on higher education enrollments. Aneree Parekh, the first author of the study summed it up as: 'Among others, we ran an analysis on the India dataset from Harvard psychologists' Project Implicit that runs worldwide "implicit association tests" that are freely accessible on their website; anyone can take them.'

'We wanted to see whether there is an implicit association of men with science and women with arts. And we did find that. It seems obvious that Indians have an implicit bias that men are more suited to science. But I was really surprised to see that so many people are also explicit about it in some surveys, with their response that Arts is more suited to women than Science.'

'So in order to hit home the implication of this idea, we decided to look at government data on enrollment in different fields of higher education in the most recent AISHE report. And we saw gendered participation in various fields of science and arts. We clearly saw that there is significantly more participation of males in science-related fields and females in arts-related fields.'

Aneree works as a psychotherapist in Mumbai and at the time of the interview, she also had a research practice at Monk Prayogshala, an independent research group committed to rigorously studying the social psychology of the Indian population. We asked her if she had any first-hand experience with biases. 'The knife that cuts the deepest is the invisible one . . . is that a saying?' she laughed as she suggested she comes across implicit biases regularly when talking about her scientific work. 'But in all seriousness, someone said to me recently: if you are a therapist as well as a researcher, how do you balance the feminine and the masculine in you? I was like . . . which one is which again?'

Another study, this one working with a large global dataset, called 'Gender Disparities in Invited Commentary Authorship' published on JAMA Network in 2019, found that relatively very low numbers of women experts are invited to write scientific articles in journals.[7] Thomas Emma G, et al. asserted:

Women in our study were about 20% less likely to author invited commentaries than men who had worked in the same field of health research for the same length of time, accumulating the same numbers of publications and citations in all key measures of scientific achievement.

'Women's ideas are not welcome or not sought. Since our dataset included more than 70,000 articles from about 2,500 medical journals, there was no doubt that this phenomenon is widespread (globally)'. Several other recent studies have seen this bias play out. These include those that looked at thousands of papers together and also those that employed AI to predict if a paper is authored by a woman or man after having studied the workings of the citation bias.[8]

This bias persists, even though there is ample evidence to show that women in science institutes are not any less scientifically productive than their male colleagues. Some top scientists have begun to challenge the bias openly. As Bengaluru-based astrophysicist Prajval Shastri put it once: 'Now it is becoming not ok to say women are less competent, many [still] think it but most won't say it.'

Yet many WILL. Indian women working in labs told us that an explicit bias against women in science is voiced by people in Indian science across the board—their own bosses, colleagues, even students, their parents, partners, harassers and bystanders of their harassment. Just as we see implicit bias sneaking up at scientific meetings, a place to watch out for explicit biases is in job interviews. As per Article 16 of India's Constitution, which ensures equal opportunities for all, no discriminatory questions should be directed at candidates at the hiring table. Yet these biases are taking place in our public science institutions.

Vandana Sharma, an atom imager at IIT Hyderabad told us about one such job interview:

'The kind of questions they ask . . . They asked me 'your husband is in the Netherlands and you are applying here, how will you manage?' I felt like giving it back to them! Why don't they realize that if I have applied for the position, I must have thought about this? Why should they worry? The interviewers must not worry about my family matters—I will take care of that. They don't ask this from any male applicants. This is really a drawback for women.'

The explicit language of the biased interview questions directly points to acceptable gender roles. Earlier generations of women interested in studying science were dismissed with words like: 'higher education will dry up your ovaries' or 'women's brains are smaller'. More recently in India, it sounds more like, as our interviewees have heard 'women come to research institutes only for rearing and breeding' or 'if everyone in your lab is a woman, who does the computation work?'. These micro-aggressions dictate wrongly that women are not interested in science and that their interests are limited only to family life. These bizarre statements are meant to police and reinforce the idea (another one we have heard spoken out loud) that 'scientific mettle is absent from the feminine mind'.

According to gender roles, to be feminine is to nurture. From babies to kitchens, hospitals, offices and labs, people expressing femininity provide most of the nurturing our social environments need. Nurture and care are inscribed in a woman's social DNA and it is what outlines the gender role of not just cis women but also other many intersex people, trans women and non-binary persons. For trans men too, the fixation on societal gender roles in science is problematic. Bittu K., a neuroscientist studying animal behaviour speaks from experiences and expertise when he says that focusing on differences in gender has real consequences that perpetuate exclusion and discrimination. 'By focusing on differences we reinforce stereotypes and biases,' he has written. As a trans man, he feels he is mostly seen by other scientists as a masculine woman and his trans identity is erased. 'This is sometimes academically rewarded (if socially disparaged) because it is misunderstood as an acceptance of the equation between masculinity and scientific ability,' he wrote in a paper.[9]

In the scientific disciplines, calcified gender roles manifest as soft and hard sciences. Femininity is relegated to soft science streams, as opposed to masculine or harder science streams. Biology is the softest in the group of 'pure and natural sciences', which also includes chemistry, physics and math, in order of perceived hardness. Further on the soft side are other care-related streams such as health, psychology, nursing and art. 'Feminine science' is poignantly illustrated in the Indian context by the discipline of Home Science, which began in 1932 as an educational path only for women students and is now an established discipline. In 2020, the UGC described Home Science while detailing the framework of the undergraduate home science degree in this way:

> Home Science is both science and social science-art related multidisciplinary field of study. The Learning Outcomes-based degree programme has been designed to integrate the application of sciences and humanities to create a cadre of home scientists to improve the quality of life of individuals, family, community and nation.[10]

As UGC described it, the discipline is meant to have the balance between art, natural science and social sciences.

Such an amalgamation is something that 'masculine' science has a lot to learn from. In home science too, the problem pointed out in its several critiques is that most 'home scientists' across the country continue to be women. Furthermore, research funding and infrastructure towards this discipline have been very sparse. Women-only colleges and universities are still its mainstays. In 2017, a proposal from the Women and Child Development Ministry aimed to improve the gender ratio by arguing that home science must be compulsory for boys in schools.[11] As we write, we still await this gender stereotype-breaking proposal to see the light of day. On its website, the Home Science Association of India states that since its inception it has 'grown in stature and has established 16 State branches and has 788 life members, 300 ordinary members, 10 Institutes and 300 student members. The association has completed 20 biennial conferences and has been instrumental in developing scientific, realistic, useful and practical job-oriented courses.'[12]

We see much potential in the new approaches to Home Science but, for now, despite the discipline's evolution it appears to continue reinforcing stereotypes about femininity and seems to propagate the notion of soft and hard sciences. In several universities and colleges, the words 'home' and 'science' awkwardly stick together.

Besides home science, teaching, rather than research, is considered suitable for women as it apparently allows more flexibility and stability. In contrast, research is redefined by the mindset as a stressful occupation. The male-dominated environment in science adds psychological pressures and frames success in science through it. Women in science have no option but to work much harder to survive, in a space where the prevailing mindset works against them and encouragement is rare. As they say in academia, 'show me the baby, not the labour pains' when referring to research outputs. Several scientists we interviewed saw clearly through the different ways implicit and explicit biases were inbuilt into Indian communities. They shared with us their realization that science and womanhood were being defined inside our institutes in a way that keeps women and other diverse groups out. A postdoctoral researcher we interviewed expressed her annoyance at the upcoming Women's Day function at her IIT. She was peeved at the session on managing work–life balance that was part of the program. 'Why is this catering to only women scientists? Do men not need a balance in work and life?' she fumed. At another IIT, a lively chemist complained while telling us of the scandal at the institute when a woman faculty member was seen smoking. 'Women must always be decent. Why?' she protested.

A powerful and erroneous scientific idea that has reinforced the gendering of a woman as a passive and caregiving figure is that of human fertilization. The male sperm is commonly imagined to infiltrate the sanctity of a passive waiting egg in the fallopian tube. Firstly, this idea conflates female physiology with 'woman'—which, in fact, is not the same thing as sex, scientifically speaking. Secondly, the logic of this scientific fairy tale is broken and indeed incomplete.[13] All of us who thought about fertilization in this way, completely forgot about the churning of a female body, cycle after cycle, to prepare an egg, the wall of the endometrium rising and falling with associated hormones and several other active physiologies that go on in the female body. Research has also shown that eggs or oocytes not only play an active part in fertilization by choosing and fixing sperms on their outer layers but also their own survival.[14] Eggs are anything but inactive. What else than a biased narrative explains why we continue to think about the egg as inactive and the sperm as the active seeker?[15]

Even the origins of the great equalizer—democracy, are marred with sidelining of women. In the very first democratic setups of ancient Greece, free men were forced to vote and women and slaves were disallowed. In India too, according to the historians of Indian science we spoke to, there were no *rishi munis* or sages that were women. In the records of

Nalanda University, a Buddhist educational complex that was functional between the fifth and twelfth centuries (and has now been revived), historians find no convincing evidence that women studied there. Likewise, in the Hindu scriptures, Dharmashastras and Varnas, the only significant 'ashram' or stage of life meant for women is the house of her husband. Gurukuls or the early schools were places where young boys from Brahmin families entered personal residencies with a guru. They lived with the guru, who fed, boarded and taught them. When finished, the graduation was marked with an elaborate Brahminical ceremony.

'The history of science IS the history of exclusion of women,' said Meera Nanda, a Professor of Philosophy and History of Science. To our knowledge, no one else has read so clearly and critically into the history of science in India as well as the rest of the world, and so deeply critiqued it. In an interview in 2021, she shared with us her realizations from studying the history of science in India: 'There isn't a single woman's name that appears in Indian history of science until the nineteenth century. Women were truly silent in the history of science and ideas in India.'

Meera has one important suggestion for all people in Indian science. She suggests: to truly resist these histories and the biases that come with them, we must first know our scientific ancestors—whoever they may be. This exercise, according to Meera, is critical to rid our collective subconscious of the patriarchal logic that justifies women's unsuitability for intellectual life.

'Evidently, those who don't remember their history end up repeating it. It is well-known that in order to understand the present, you got to know where you're coming from. More than that, I am convinced that those ideologies, those stories that justified that a woman's place is at home and not in science have not lost their power yet. I think patriarchal stories and myths are very much part of our collective unconscious,' she said.

In many parts of the world including our own 'the active component of nature is given a male valency and whatever is considered passive is given a female valency. And activity is preferred over passivity. So gender gets built into natural philosophy itself.' Natural philosophy is, of course, the archaic name for science.

Courtesy: Gregor Reisch/Public domain, Wikimedia commons

Allegory 'Astronomia' in *Margarita Philosophica*, 1503.

Meera provided a timeline of women's historic exclusion in a lecture at an astrophysics institute in Pune in 2019.[16] She did so with the help of historic paintings, sketches, early encyclopedias and myths. Pointing at a sketch from one of the earliest compilations of general knowledge, the *Margarita Philosophica* from 1503 she said: 'Astronomers will recognize this figure to the right, this is Ptolemy, and this here . . .' she pointed at the female figure on the left labelled Astronomia ' . . . is the muse of astronomy.'

Going to the next slide she said: 'This is Johannes Hevellius [yet another early astronomer] on the front cover of his Star Atlas presenting his book to, again, the muse of astronomy Urania. And in the background, surrounding Urania are all the major astronomers of that era. They are all included from Ptolemy onwards through Galileo to Johannes in the seventeenth century.'

The front cover of Johannes Hevellius's Star Atlas dated 1687, but not published until 1690.

Meera then laughed as she said: 'The muse of astronomy is always presented as female even though there is no female to be seen anywhere amongst them . . . the fathers of science.

An extract from the frontispiece of the Encyclopédie (1772). The work is laden with symbolism: the figure in the centre represents truth—surrounded by bright light (the central symbol of the Enlightenment).

The muse remains female. A single woman participates in this whole process.'

In the image on Meera's next slide was another female muse—this time a representation of Truth.

'This is the encyclopedia of the enlightenment age from the eighteenth century. The woman figure in the centre is *The Truth*. And she has a veil on her face, which is being removed by *Reason*, there is one male figure and it represents reason. And around them, if you look at it closely, are all female figures representing geometry, botany, zoology, art and sciences.'

Through the three slides, Meera pointed out the contradiction in the exclusion of actual women interested in science in those times and the gendered reading of nature as a woman. In reality, there were no women represented in scientific societies and universities at the time when these images became banners of science. 'But in imagination, in ideology, knowledge was projected as feminine,' she said.

Meera can't get her head around why nature and science were being idolized as goddesses. Her guess is that it was probably one way of compensating or some kind of unconscious storytelling. 'I'm sure they could see the injustice of what was going on. And this was probably one way to make it seem all right.' For the same reason, Meera is certain that goddesses cannot be a feminist symbol. 'It's not just in the West. In Hinduism, it is the same story. Saraswathi is one of the many goddesses we worship, another Shakti which is a form of Prakriti, or nature, emerges from the body of Shiva as a goddess.'

'I tend to have a mental dialogue going on with some of the cultural feminists in the West and in India, who have turned to the old goddess images as a source of empowerment. They are so sick and tired of patriarchy around them, they find the image of Kaali or Gaia empowering. They are looking for new symbols. I argue that goddesses represent the unconscious overcompensation toward keeping women in their place and symbolically elevating them, while materially suppressing them. I have come to the view that there is an inverse relationship between the worship of goddesses and empowerment.'

Scientifically minded women working in India today told us about the bias from their own lived experiences. From their accounts and from Meera's reading of women's exclusion in the history of science we understand that the bias against women in science is a part of our cultural knowledge.

'The sun does not discriminate,' astrophysicist Meghnad Saha once said while protesting caste-based gatekeeping he observed in Indian science and pointing out that knowledge is not his or hers.[17] Yet, the frameworks in which we engage with science are always mediated by our cultural knowledge. Social realities may not totally determine what we can discover but they play a big part in how we think about nature. Addressing this invisible bias is truly an important part of the reparations to be made in the hope of closing the gap and changing the way we see science and nature today.

12

The Perfect Conditions for Sexual Harassment

In October 2016, IIT Delhi invited us for our very first opportunity to talk about the lab-hopping project. Since that first time, delivering talks at the same institutes we hop through has been complicated. Each time, we expect to discuss our work, critique it and, in the best case, receive a few pats of encouragement. But often, it ended up being unexpectedly painful to repeat the horrendous stories to the same rooms they had come from.

Extending from our talk at IIT Delhi, the organizers arranged for a panel of women faculty members to share the stage and openly speak about their experiences. Journeys and survival strategies were shared, along with much appreciation for one another. The feel-good sisterhood dissipated quickly when an audience member, a postdoctoral researcher at IIT Delhi, posed the question: 'What am I supposed to do when I am whistled at on campus?' The panel responded with disbelief. They shook their heads. IIT Delhi was safe, they corrected her, unlike the rape capital that lies outside the campus gates. 'Such a thing would never happen,' the panel was heard saying. This statement stoked a contrarian deluge of incidents from young women scholars in the audience. One after another, we heard of multiple incidents of sexual harassment that had taken place in and around the institute's labs, lecture halls and living spaces.

Soon we were out of time but by then, the devil was out. Each of us walked away with some discomfort in the realization that sexual harassment is a taboo topic within the science community. Since then, we've noticed that on the odd chance of the topic of sexual harassment being brought up, the discussions quickly get entangled with the complicated workings (and failings) of due process. It is a gridlock that prevents any relief for the many victims of sexual harassment in our community, and a meaningful discussion on the topic remains out of reach. One common question continues to exasperate: Why are our institutes unable to address sexual harassment?

The claim that our science labs are unsafe spaces is voiced by a large number of sources, ranging from anecdotal evidence and media reports to large-scale studies and commentary from scientists. 'Late-night research, isolation and a strict, male-dominated hierarchy are the perfect conditions for sexual harassment, say students at IISc,' reads one 2018 report in *The New Indian Express* titled 'Forced to Work under Sexual Harassment for 5 Years IISc Student Reveals Truth About Institute'.[1] In our own investigations, we found that young women in science echo this point of view. In the same year, the existence of these 'perfect conditions' was also acknowledged in an open letter by 165 Indian scientists addressing sexual harassment in the community. Urging the wider community to support multiple victims of a high-profile sexual harassment case, they hinted at the problematic hierarchical dynamic between students and PIs in their open statement: 'By no imagination is sexual harassment at the workplace exclusive to the world of scientific research, yet some features of how science is organized, makes its authority structure especially perilous for women,' they wrote.[2] This analysis is attested elsewhere. In many cases of sexual harassment in Indian science and on a #MeToo form we set up on thelifeofscience.com, the victims are graduate students. This power dynamic leaning towards the PI makes the nature of harassment severe. One of the respondents to our #MeToo form, a PhD student and victim of sexual harassment at Jadavpur University, often heard from her abusive advisor: *'Five bochhor er jonno tui aamar'*, which translates to: 'For the next five years, you are mine.'

In 2018, a report titled 'Sexual Harassment of Women Climate, Culture, and Consequences in Academic Sciences, Engineering, and Medicine' was published by NASEM, one of USA's science academies. It reviewed decades of global research on sexual harassment in STEM as well as conducted its own original sociological research. The authors wrote in the report:

> Through our work it became clear that sexual harassment is a serious issue for women at all levels in academic science, engineering, and medicine, and that these fields share characteristics that create conditions that make harassment more likely to occur. Such environments can silence and limit the career opportunities in the short and long terms for both the targets of the sexual harassment and the bystanders—with at least some leaving their field. The consequence of this is a significant and costly loss of talent in science, engineering, and medicine.[3]

The vast body of research they reviewed and conducted also showed that the horror of sexual harassment can be kept in check if institutions and those leading them simply wake up to it and start listening to the victims.

The NASEM study said:

> We are encouraged by the research that suggests that . . . institutions can take concrete steps to reduce sexual harassment by making systemwide changes that demonstrate how seriously they take this issue and that reflect that they are listening to those who courageously speak up to report their sexual harassment experiences.

Indian science academies or institutions hardly ever bring up sexual harassment, except for the odd nod to studies like the one discussed above.[4] An in-depth and prescriptive study of their own is much-needed but hard to come by. Nor have our institutions paid heed to the necessary mitigative steps that have been suggested and implemented elsewhere. As a result, sexual harassment is rife in Indian labs and its tolerance continues. Smaller groups of scientists have, indeed, come together to express the urgent need for publicly available data on sexual harassment in institutes, since, without it, they are aware the system will not budge. For example, women members of the Astronomical Society of India (ASI) have stated that India urgently requires statistics detailing the number of sexual harassment cases institutes have dealt with, how many institutes and universities have functioning Internal Complaints Committees (ICC), the quantum of punishment recommended by existing ICCs, how many decisions have been taken in favour of accused, how many were dismissed and how strict is the implementation level of mandatory procedures at ICCs. In their discussions, Indian women physicists observed that most students did not know the names of people in their own complaint committees. They noted that guidelines from the UGC, the statutory body in charge of higher education in India, discuss all safeguards for victims but institutes are failing in implementing them.[5]

While some top scientists in India have been swift in expressing their grave concern over sexual harassment in labs in public events, conferences and editorials, they fall short when it comes to holding their own institutes and their mechanisms to account. We have heard established scientists, in public and private fora, concede that sexual harassment is an impossible problem to tackle. A senior scientist at a central university once said at a forum: 'Nothing can be done about sexual harassment in science institutes. Victims should go to the police.' The implication of this statement is that due process is defunct. It's no wonder then that most victims of sexual harassment in Indian science do not bother to file official complaints.

Suffering women have no option but to seek different modes of resolution, and for the more privileged of them, an opportunity came in 2018 with the #MeToo movement when hordes of women in India took to social media naming their harassers and publicly speaking about their experiences. Though India's #MeToo movement featured primarily urban upper-caste and upper-class women from the media and journalism, the event that sparked it off was a much more grassroots initiative from Indian academia. It was a list by a law student Raya Sarkar. Raya's openly accessible 'list of sexual harassers in academics' (LoSHA) featured names of academicians accused of sexual harassment. Most of the perpetrators' names were from the social sciences and one was from the natural sciences.[6]

Noting that the Indian science community was relatively silent during the movement, we created a new closed form targeted at it in 2018. The form aimed at collecting and responsibly reporting cases of sexual harassment in Indian STEM, while supporting the victims in any way they needed. Compared to the extent of harassment in Indian science that we have come across during the lab-hopping exercise, the responses we received were far fewer. We are choosing to not share details of incidents simply because most of the respondents did not want that. Among the entries we received from targets of sexual harassment, unsurprisingly, most chose to stay anonymous; interestingly, many did not name their harassers either. At first, we presumed this may be because the women are aware that exposing their perpetrators' names could put them at risk. Even with their own names hidden, powerful perpetrators could identify them easily even with small indicators, and drunk on the impunity afforded by the boys' clubs, continue to jeopardize their careers. But this was not the complete story. A respondent to the form enlightened us further:

> I wish to clarify one thing that the ladies who join the job of Research Assistant and Senior Research Fellow in research and educational organizations are more prone to sexual predation because they are very helpless and their social background and financial strength don't allow them to narrate their past in the #MeToo like campaign. Definitely, there is a social blockage in the form of stigma linked to the strata of the society from which we came from and later settled. Therefore, we can't be compared with the group associated with the entertainment industry, corporate and journalism because these ladies are far more vocal, psychologically strong, expressive, independent, and financially very strong and recognized in the society.

As we closed for entries in early 2022, the #MeToo form had trickled in close to 50 accounts of incidents of sexual harassment in the Indian science community. The perpetrators were reported to be the colleagues or professional seniors of the victims in almost all cases. The incidents took place in institutes across the country, from cities such as Chennai, Dehradun, Bengaluru, Gandhinagar, Pondicherry, Kanpur, Mohali and many others.

One of the questions in the form was: Have you sent out an informal or formal complaint, to whom and how was it dealt with? The answers to this question are illuminating because they make apparent the state of 'due process' in Indian academia. The few who did opt for due process have expressed their dissatisfaction and frustration with the handling of their case. 'I was allotted a different lab. He was allowed to stay on the job until his contract gets over and even allowed to take more girl students. He is still on the job,' said one respondent. Another said that her official complaint was dealt with lightly and hushed up. One respondent was simply asked to go to the police, something she chose not to do being new to the city, bogged down by health issues and moreover having 'no support whatsoever from my guide'. In many cases, the victims chose to first share their grievances with a known figure like a professor or a colleague. Sadly, this person either ignored them or advised them to desist from formally complaining and, in one case, turned hostile. One was scolded for not being able to 'take compliments' and another wrote in vain to the Dean and the President since she did not trust the ICC head. One of the respondents was too close to the end of her degree to risk losing it all, another was informed she 'could not complain' since she had completed her project. Then there is the one respondent, who was a newcomer at the time she was harassed. Presumably unprepared to cause a stir, 'I quit within 1.5 months of joining,' she stated.

One answer to the question summed up the broad conditions facing most survivors: 'No, the environment within the institute was not conducive for a fair process of the complaint.'

India's laws mandate the presence of an ICC to tackle complaints of sexual harassment in every workplace. Unfortunately, in almost every sexual harassment case we heard and read about, women in science report improper ICC procedures. This is exacerbated by a generous serving of victim blaming and slut and body shaming taking place inside our institutes. People serving in these committees, even when they are women, often look the other way especially if the accused is a fellow faculty member. The lack of bystander training has been perpetuating all the creepy gender dynamics playing out in our labs. Within tight-knit circles of power, it is incredibly rare to see scientists stand up for victims when the accused is a friend or respected colleague. There is also a belief that one must be wary of believing victims as their stories may be cooked up with the aim to pull down successful men of science or as an act of vengeance for whatever reason one can imagine. Patriarchy is so deeply rooted in the community that the sexual harassment narrative is more likely to be seen from the lens of the damage it can do to the career of the accused rather than the trauma it has caused to the survivor. The culture of impunity is affirmed with every case that is unreported, discarded or settled, and those that end merely with resignations or warnings. In rare cases where the accused is found guilty by institutional bodies, he is sent into 'forced retirement'. The language of the law enables institutions to centre the well-being of the perpetrator as it prioritizes causing him minimum damage. In a 2021 case against an IIT student, the state High Court freed the rape-accused student while referring to him as 'talented and the state's future asset'. This prompted the survivor, his peer at IIT to respond in a media interview: 'If the court is deciding [based] on the fact that he is an IITian, then I am also an IITian.'[7]

In the same report, she opened up about how isolated she became during the tormenting episode.

> There were a lot of people openly defaming me with my name being mentioned. Even the people who helped me reach the hospital have been threatened. The ones who used to visit me at the hospital were threatened by some seniors. They stopped visiting me.

It seems top institutes set a precedent for the rest of India's science ecosystem—both in its high quality of research as well as the bad standards of checks and balances with regard to sexual harassment. Currently, the bar is very low.

In another case at a premier institute in south India, Latha* was backed up into a wall and threatened by a male colleague during a professional meeting. 'I couldn't believe what had just happened,' Latha told us. 'My brain stopped. His tummy was so close to me . . . I could see my colleague nearby shivering. Even a student happened to be in the room, doing some work. The next moment everyone behaved as if nothing happened.'

Latha's struggle spanned years and included both due process and a police case. Despite there being many witnesses, she received very little support from bystanders at her institute. She said in an interview, 'It is very interesting, the moment there is a fight with the man, immediately everyone is looking at my publications. Very openly, many professors say that the problem is that he has many publications, unlike you. This shows how our institutes are reasoning . . . if you have so many publications, you are worthy enough to harass or misbehave with anybody.'

In Latha's case, her harasser was merely asked to apologize and only received a censure. He has since then been promoted, received salary hikes and continues to be 'on every committee—technical, cultural, convocation . . .' Latha, meanwhile, is struggling with long delays on her own promotion. She admitted to being mentally and physically drained. 'I have become very low. Thanks to him, the whole institute knows I'm divorced and remarried—something I don't mind sharing but why should everyone know? People shift their faces away when I walk by,' Latha said, her voice slightly breaking.

A simple online search of the keywords 'vice chancellor India sexual harassment' shows how pervasive this issue is in the highest echelons of academia. 'Sexual harassment case: Former Panjab University V-C served legal notice for making committee proceedings public' on *Hindustan Times* (2020). 'Amid #MeToo movement, sexual harassment-accused DU Professor Bidyut Chakrabarty appointed V-C of Visva Bharati University' on firstpost.com (2018). 'Minister & secretary disagree but govt extends tenure of V-C accused of sexual harassment' on theprint.in (2021) are only some of the headlines to be read. Often, dealings of these cases at high levels are politically motivated.

When our collective showed solidarity towards a former PhD scholar who had publicly outed a professor as a sexual abuser in a YouTube video, we received multiple angry emails. The emails were from his colleagues (at least one of whom we later found out was related to him) and even students, who believed that any accusations against this professor were bound to be politically motivated since he was well known for his anti-establishment stances. Somehow, they were certain that the woman in the YouTube video was lying. At the same time, we also began hearing from others at the professor's institute that he was known for his inappropriate behaviour and relationships with students. The details were confounding but what stood out was the way supporters of influential men in science are capable of silencing any potential victim. This is a common theme among the sexual harassment cases we have investigated in depth. In another case, students were reported by the *Indian Express* to be stressed due to the 'constant presence' of a sexual harassment-accused biologist at their university. The professor had returned to his lab on a clean chit after getting arrested based on the complaints of multiple students accusing him of sexual harassment. This professor went on to enjoy full impunity and continues to be celebrated for his science by some media channels.[8]

Consequently, the uncomfortable reality of Indian science being rife with sexual harassment is covered up. Seasoned sexual predators armoured with social and political capital have been getting away scot-free. We've had the privilege of reporting just one case, where it can be said that justice was served. The case filed by Swati* against her department leader Shyam* at the Indian National Archaeometallurgy Institute in Bhopal closed after almost a year-long internal due process with the sacking of Shyam. The reputed scientist was found guilty of sexually harassing Swati for a number of years. We feel it is important to tell this story since it includes repeating patterns that are common among the many sexual harassment cases and at the same time it is a rare case where the victim was properly heard. Importantly, we can share a redacted part of Swati's story since she is prepared to tell it.

As Swati began to speak about the ordeal at the interview we had arranged one chilly night in 2020 at a cafe near her home, a shared feeling sprung up between the three of us that a wound was being opened. She put us at ease by saying:

'Look, I fought a case to teach him a lesson and ensure he doesn't abuse any other women. I seriously don't think of myself as a victim because I played my part and he was fired in the end.' Thankfully, Swati succeeded in putting an end to the cycle of abuse splayed out by Shyam targeting not one, but many young women scientists, spanning several years. Repeat offenders are aplenty in Indian science. The whisper networks as well as the responses on our #MeToo form attest to this fact.

'It all started with good morning messages and slowly but eventually got worse,' Swati shared. Somehow in the interest of their survival in academia, victims learn to deal with the harassing comments and Swati did the same. Until the harassment peaked one evening in an incident of attempted sexual assault. After this, Swati started to be extremely cautious. But Shyam persisted. Thereafter began a cycle of abuse, far too common in professional spaces, where the predator retreats after being confronted and readjusts his behaviour for a short duration of time, only to eventually restart his advances.

It was after a workshop at their institute that Swati decided that she would file an official complaint. The workshop's purpose was to inform staff, scientists and students about the ICC. She shared her desire to file a complaint with another woman colleague in her lab, who then confessed that she had also been at the receiving end of sexual harassment by Shyam and there were many other victims. 'He has a history that has been covered up. I decided that I will try to put an end to his abuse.'

The next day Swati went over to see the Director of the institute. This is where the rarity of this case begins: in the way it was correctly handled. The Director listened intently and laid out all of her options, ranging from switching departments to going to the police with the complaint. 'I told the Director, if you only change my department, what will happen to him? There will be no punishment for him.' So Swati then sent a formal email to the Director, spelling out the harassment that had taken place along with the statement that she wanted to file a case in front of the ICC. 'The case was transferred to the chairperson of the committee and from there on, it actually started,' Swati said.

The first thing Swati asked of the committee was a request to suspend Shyam from the premises as prescribed by the Prevention of Sexual Harassment Act to protect the case and the potential victim. Once, she felt safe she could gather and provide all the evidence she had. Swati credits her eventual success to the evidence she had collected over the years of abuse.

'I would have definitely lost this case if I hadn't saved those emails and the messages. To those who are considering filing a sexual harassment case, I want to advise you to, if you can, wait to collect evidence first. Without the evidence, the best we can hope for is a warning to the harasser. And nothing stops that person from repeating his offences. So I say to everyone and anyone: always record it. Do it! Play it smart!'

Swati was satisfied with the composition of the ICC at her institute. As per the law, the ICC must consist of members from inside the institute as well as outside; it must also include a lawyer, if possible, and a social worker.[9] The complainant is free to seek legal help to defend her case, which Swati did.

'Altogether, in my case, the committee played a very big role. I know in India, mostly the reverse happens. They went by the facts, and they went by the evidence I presented. I have to say—they were neutral. This is good enough for me because my case was clear. The only thing I wanted was for them to play fair, because I knew that if they play fair, then I will be successful.'

According to official guidelines, a complete redressal must be provided within three months of filing the complaint. But even Swati's case, one that we are considering relatively successful, lasted close to one year. Shyam had also tried to informally settle the case, an illegal move that went against him in the final judgement. The decision was made, as it must be, in four steps. First, the institute ICC committee recommended its judgement after rounds of investigations based on the evidence provided by both parties. The ICC's judgement recommendation was cleared by an external committee— made up of independent individuals within the Indian science community and then by the Director of the institute. The final judgement needed to be approved by a government department that acts as the board of the institute. The decision was taken to terminate Shyam's employment at the institute. This is the strictest decision an ICC can take as a quasi-legal body. Beyond this, the victims of sexual harassment can file a criminal complaint with the police who are governed by laws under the Indian Penal Code. Like most survivors, Swati chose not to do this.

It has to be noted, however, that Shyam still enjoys the courtesy and patronage of several scientific circles in the country.

'The science community needs to boycott these kinds of people,' said Swati. 'You need to remove them from advisories and committees. Only if such people are removed from the community, progress can happen. What's the point if he is fired but his friends are always there for him? I say we need to remove them as well. We need to stand up against sexual harassment no matter what. This is what we want.'

Listening to the survivors of sexual harassment is the one minimal step our scientific institutes need to take immediately. It is owed to all those who have taken the risks to go through with the due process and also to those who are being silenced as we speak. It has been really difficult to put the above together as we went back and forth with sources and tread very carefully over long-lasting traumas and risks to what remains in the aftermath. And what we are able to bring on record is only the tip of the iceberg; most horrendous realities remain hidden behind the hushing curtain of our science community that chooses to look away.

13

Hiring Bias

Kamala Sohonie graduated from Bombay University in 1933, at the top of her class. She followed her father and uncle's footsteps to continue her education at their alma mater, the prestigious IISc in Bengaluru. However, unlike her father and uncle, Kamala's request for admission was dismissed. C. V. Raman, the then-Director, according to the book *Dispersed Radiance*, proclaimed 'I am not going to take any girls in my institute.' What happened next is history. 21-year-old Kamala, instead of retreating, confronted the revered Nobel laureate and staged a protest outside his office until he relented, albeit with conditions.

How much have things changed at the hiring level since 1933? It depends on whom you ask.

Courtesy: Anil Sohonie

Kamala Sohonie, a renowned biochemist, once studied at IISc.

In 2015, two professors from Cornell University created a flutter with the results of their survey. Wendy M. Williams and Stephen J. Ceci used their survey to conclude that gender bias in hiring no longer exists in the US; not only that, women actually had an edge.[1] Employers, according to the survey, were twice as likely to prefer women to men. The paper said that the greatest harm to women considering careers in academic science was not discrimination, but their own decision to not apply. It concluded:

> Efforts to combat formerly widespread sexism in hiring appear to have succeeded . . . The perception that STEM fields continue to be inhospitable male bastions can become self-reinforcing by discouraging female applicants, thus contributing to continued underrepresentation, which in turn may obscure underlying attitudinal changes.

According to Williams and Ceci, the playing field had successfully been levelled.

These claims were widely reported by the likes of *Nature News* and *The Economist*. And why not? The paper was stating something positive that seemed to be backed by data. Moreover, if true, this also meant employers no longer needed to struggle for gender equity; after all, they were not to blame for the gender gap. In an opinion piece that followed the publication, Williams and Ceci said:

> The low numbers of women in math-based fields of science do not result from sexist hiring but rather from women's lower rates of choosing to enter math-based fields in the first place, due to sex differences in preferred careers and perhaps to lack of female role models and mentors.[2]

Results like these are just the evidence needed by scientific employers to support the dominant perception that institutions are already doing as much as they can to hire women. And this is, in fact, the impression we get when we speak to authorities of Indian science. If we investigated the gender gap by speaking only to these big bosses, we may have concluded that the times of hiring bias have passed, that bigoted C. V. Raman-esque attitudes to women who apply for scientific jobs no longer exist, and that the gender gap is the result of women's own choices. Therefore, there is no point in haranguing employers in the name of gender equality. Instead, let's be positive, we might have advised. Let's accept that the battle is won, at least from the systemic side of things and shift the focus on fixing the women and their mindsets. Or even better, let's just accept the gender gap in science as an outcome of 'sex differences'. All is well.

However, early on in our lab-hopping journey, we decided that we would not be interviewing just the occupants of the ivory towers. We knew that an honest picture of Indian science could only be painted through the experiences of the 'everyday' scientist. And these experiences made it clear to us that gender bias in hiring persists, and it has taken on various insidious avatars.

It did not take long for gaps in the Williams–Ceci study to be exposed. In subsequent commentaries, it was accused of having fallen prey to the 'superstar problem', a term coined by American feminist legal scholar Joan C. Williams and psychologist Jessi L. Smith. In their response to the study, they argued that equality for 'superstar' women scientists was not the same as equality for all women.[3] It may be that women scientists with Ivy League degrees, with publications in *Nature* and *Science* and women Nobel laureates are treated on par with men, but what about the others—those women in science who do good research but don't have such exalted credentials. Are they being hired equally as their male counterparts? The stories we heard from the ground in India suggest overwhelmingly that the answer is no.

Those running our scientific institutions would have us believe that there is nothing guiding their decisions apart from the pursuit of excellence. If this is true, one may very well wonder why Indian science is still underperforming. Even much-revered scientists like Bharat Ratna awardee C. N. R. Rao have lamented that Indian laboratories are 'rife with mediocrity'.[4] This raises the question of whether employers truly are hiring based on excellence alone. And if so, who defines what being excellent is? Are these standards for hiring uniform across all axes, such as gender and caste? Our conversations have suggested that to be hired, women are expected to have not just top-notch subject knowledge, but also confidence, connections and an agreeable personality. These are the traits that make a potential candidate a 'superstar' in the eyes of an employer. When they detect such a superstar, it may indeed be that the person is welcomed and treated very well, irrespective of their gender. Men, however, seem to be judged more leniently. Defects in personality ranging from rudeness to predatory behaviour are tolerated (sometimes even glorified) if their science is good. Even faulty science may be excused if he has the right connections and talks the talk.

The question is this: are as many mediocre women being hired as mediocre men? In the words of American feminist Bella Abzug: 'We don't want so much to see a female Einstein become an assistant professor. We want a woman *schlemiel* to get promoted as quickly as a male *schlemiel*.'

We repeatedly encountered evidence of a double standard during hiring but it hit us most starkly on one bright day on the west coast of the country in 2019. We had multiple interviews lined up and the first was with Meenakshi,* a senior scientist who was the Director of Research at her university. After regaling us with stories of her accomplishments, she added that she has felt treated better in India than abroad. She said, 'Here in India, women don't have to be equal, just 75 per cent as good. If a man and a woman are equally good for a job or award, the woman will be selected. We should get rid of this idea that women don't get awards. Maybe it's time to introspect and think that someone else is better.'

Such questioning of the merit of women is something that is implied very often but most bosses these days avoid saying it out loud for fear of sounding politically incorrect. We were intrigued to meet a woman scientist herself who believed that women were treated better in science than men. A few minutes later, we were further taken aback to hear Meenakshi add:

'Sometimes, if there is a position for which a guy and a girl apply, I'll take the guy. Because I already have so many women. They will have to run away at 4.30 p.m. to catch a bus and go home because the baby is waiting. If it's a man, I can tell them to stay back and finish up the work. In India, being a lady comes with a lot of responsibilities. Women are good to hire for nine-to-four jobs but for research . . . you know, I never used to go home before 9 p.m. It's a research institute so I have to look at it this way.'

The belief that men are more suitable for research is sadly common among many bosses of Indian science. They are able to insist that women have it easy, even whilst explicitly discriminating against them. We found Meenakshi's views jarring and tried our best to make sense of it. We decided that perhaps she was fooled by her own experience as a superstar scientist. Having been treated well most of her scientific life thanks to her own accomplishments and her academic pedigree, she had wrongly assumed that all women have the same experience. She was misconstruing equality among superstars as equality for all like Ceci–Williams, while actively discriminating against them herself.

The more scientific spaces we visited, the more young women in science we met. And the more conversations we had with them, the clearer it became that gender bias in hiring still exists. At conferences, the sight of scores of smart and confident young scientists arriving from their international PhD and postdoctoral laboratories always made us a tad anxious. All these young people were seeking a faculty position in their home country. They knew as well as we did that only a small percentage of them would be successful. Though the percentage of men and women among PhDs is comparable, institutes claim to have a harder time finding worthy women to hire than men. For whatever reason, far fewer women are judged good enough to hire.

So, what exactly happens in this time of 'judging'? Prabha*, a physicist at a government research institute, shared what happened behind the scenes when she began the process of hiring her first PhD student. The interview committee was headed by a senior scientist who was an ex-HOD of Prabha's department.

'We had five or six applicants to consider for the position. All of them were women, except one. The chair, a male, looked at one CV and said "oh she is married". I pointed out that this did not matter and we should at least call her for an interview. Then he picked up the next CV, that of an unmarried woman. To that he said "oh she is not married, but she will get married along the way" and he threw away the application.'

Prabha was livid and refused to accept this, as this applicant had a very high score in the qualifying exam. Eventually, the applicant did very well at the interview too. 'The male applicant also came for an interview and he was terrible. Yet, the committee chair asked me to take the boy. I said no, I will take the girl who really performed well. That is only fair. If she doesn't join, I will advertise again.'

The committee chair agreed though he wasn't very happy. He left Prabha with a prediction that the woman would leave her in a month. 'She got married in between but I didn't care as long as the work was done. Today, she has finished her PhD, published two papers and got a job as a forensic scientist.'

'This is what women students face every time!' said Prabha. 'I wanted to remind the chair that he had a daughter. She might have to face the same thing tomorrow.'

The gender bias at hiring is not always as upfront as what Prabha witnessed. They manifest in much more insidious ways. One example of this is the unofficial ban on hiring couples, which selectively discriminates against women scientists. Another is age limits.

Few things make Farah Ishtiaq's blood boil more than the mention of age limits. An ornithologist and disease ecologist, possibly the only one in south Asia to be studying malaria in birds, Farah was haunted by the baggage of her age for many years. She came back to India in 2012 after nearly a decade of postdoctoral research experience in the UK and the US as well as a number of prestigious fellowships to her name. Having worked under big names such as Pamela Rasmussen, an iconic expert on Asian birds, Farah expected that her impressive portfolio would hold her in good stead during her job hunt in her home country. However, Farah was unable to secure a faculty position till 2019. The reason? She was over 35 years old.

Being 35 or under is an important criterion for faculty hiring in Indian academia. Though age limits are not officially mandated, most job advertisements for entry-level 'Assistant Professor' faculty positions state that the applicant must be below 35. The understanding is that younger faculty members have more time to achieve superstardom. For reasons perhaps reminiscent of why younger brides are preferred by most Indian families, younger scientists are thought to be more productive and adaptable to the work culture at the institute. Most importantly, younger scientists stand the best chance at winning awards, such as the Shanti Swarup Bhatnagar Prize, which has an age limit of 45 years. Nothing raises the prestige of an institute like being the home of Bhatnagar Prize winners.

Like many women in India who navigate personal responsibilities on the side, Farah's academic trajectory took longer than what is considered ideal. When we caught up with her in 2019, she spoke to us about graduating with a PhD from Aligarh Muslim University at 27 and following that up with four years of non-academic work at the well-known wildlife NGO Bombay Natural History Society. Only then did she resume postdoctoral work. Farah was 38 years old when she applied for a faculty position at the prestigious Centre for Ecological Sciences (CES) at IISc Bengaluru. She was shortlisted for an interview, which she recalls went very well.

'The then chairman of the department was very positive about the work I was doing and I was called for a second interview. He told me to apply for a couple of fellowships and if I got them, CES would host me. I was awarded both the fellowships.'

This meant that Farah was invited to set up a lab in IISc and do her work using the grant money but only temporarily. If CES did not offer her a faculty position in five years, she would need to look for a new host institution. 'Still, I thought this could work. I had hope,' she said. Funds and publications trickled in and Farah's research bloomed, yet there was a problem. Farah's hope for a tenure-track position at IISc was withering away. 'I could not get PhD students for myself because for them it was a risk—there was no guarantee how long I would be here.'

At 46, Farah was met with more and more resistance to being hired, not just in IISc but in other places too. So much so that she prepared herself and her family to leave the convenience of Bengaluru if needed. She would need to cast her net wider for any chance at a permanent job. In a system that is too steeped in inertia to tweak arbitrary guidelines, Farah found herself unhireable despite being qualified—at that point, the problem was that she was overqualified. It is rare for a scientist to be hired at the level of 'Associate Professor', as most institutes are expected to give preference to promoting the Assistant Professors already on their payroll. Knowing this, Farah made peace with the possibility that she may have to settle for an Assistant Professor role. 'Beggars can't be choosers,' she said. Still, for a long time, nothing worked out for reasons incomprehensible to her.

'At one place where I was invited, the chairman said I was overqualified for the entry-level assistant professor position and that they would not be able to hire someone at a senior level. If they'd made up their mind, why did they call me?! Saying "age doesn't matter" is a sham.'

It was only in 2019 that Farah saw light at the end of the tunnel. She joined a non-profit research institution as a senior scientist. Meanwhile, she continues to speak up about the way the system fails scientists like her due to its obsession with age.

The age limit debate crops up now and then on Twitter, which in the past few years, has emerged as a popular platform for discussions about the Indian science community. In 2019, K. VijayRaghavan, who was then the Principal Scientific Adviser to the Prime Minister of India, tweeted that institutions, rather than the government, had to take the onus to remove biases at hiring:

> All these issues should and are best addressed at the institutional level, where they have been created and where there is context-dependent flexibility. Asking the government to step is to ask for a one-size solution. Institutional leadership should grasp the nettles and not stare at them.

He encouraged institutions like IISc and TIFR to quickly make formal exceptions for women and hire older people at higher levels. The government would not stop them from doing this, he confirmed.

A case of strong institutional leadership took place at CES, interestingly the same place where Farah was struggling to gain a permanent post. Ecologist Maria Thaker had just found out that her application for tenure had been rejected. Maria knew that this was because the all-male panel who assessed her did not consider her six-month maternity leave. Subsequently, she was compared with a male colleague who had more publications. The result was that she received a three-year contract extension instead of tenure. She would need to repeat the procedure all over again. Frustrated, Maria was pondering over her options when she had a chance encounter with senior biologist Sandhya Visweswaraiah. Sandhya took up Maria's cause and successfully lobbied for a change in IISc's tenure policy. Since then, the tenure clock of women researchers at IISc is paused by one year for every child. As a result, when women faculty at IISc apply for tenure, they can do so on the back of publications in the last six years (in case she had one child) or seven years (in case they had two children). Sandhya explained in an interview, 'It was also mandated that when they do get tenure, it is back-effected [with respect to pay rise and position], so it is not that you are one year behind the male colleague. Just because you give her tenure later, don't put her one year behind a male colleague who joined at the same time.'

Age limits are detrimental not only to married women with children. Our conversations with women with intersecting marginalizations highlight that this is a grave oversight. Rupa*, a physically disabled scientist, opened up to us about her struggle to get promoted. Despite teaching for 20 years, she was still only an Assistant Professor. According to her, changing rules and technicalities were holding her back, while her male juniors had all gotten promotions before her.

Shalini Mahadev, a researcher in neuroscience at a central university, pointed out to us how deeply set back she is by the age limit. If science is mentally, emotionally and physically taxing for the average woman researcher, it is even more so for those like Shalini, who belong to marginalized castes. Added stresses, often unrelated to marriage and motherhood, and the lack of social advantages cause many students of science to start their PhD late and take longer to complete it. 'Right now, it's ageism that affects me the most,' said Shalini during a virtual chat. 'Though I have many years of expertise in science, I'm constantly worrying that I'm 37, I don't have papers, I'm not getting a job,' she said. Policies like the one that helped Maria at IISc will not help Rupa and Shalini.

Without introspection, the Indian science community is being fooled into interpreting the successes of some women scientists as gender equality. The question begs to be asked: when the choice is between an 'average' male's CV and an 'average' female's CV, how do gender, caste, disability, etc., bias this decision? Studies from the West have well established that race and gender do play a role in hiring. For example, a 2019 study in the journal *Sex Roles* showed that Bradley Miller (a common white male name) is more likely to be hired than José Rodriguez (a common Latino male name), Zhang Wei (an Asian male name) is more competent than Jamal Banks (commonly interpreted as a black male name). And both Miller and Zhang are more competent and hirable than Maria Rodriguez (common Latino female name) or Shanice Banks (black female name).[5] Similarly, a popular 2012 study published in the PNAS journal showed that biology, chemistry, and physics faculty members preferred to hire 'John' for a lab manager position, even though John's application was identical to 'Jennifer'.[6]

A similar study was conducted in India over a decade ago but somehow has not caught the imagination of Indians as much as the American John–Jennifer one. Economist and former chairman of UGC Sukhadeo Thorat conducted a study in 2007 to examine the prevalence of discrimination by private companies during the hiring of Dalit and Muslim candidates. The authors replied to job advertisements that appeared in English newspapers with three applications each. The three were identical except that in one they posed as an upper-caste Hindu applicant, and the other two as a Dalit person and a Muslim person. Upon statistical analysis of the resulting data, they found that it was 33 per cent less likely for the Dalit person to be invited for an interview and 66 per cent less likely that the Muslim person would be called.[7]

If indeed the CV of a below-35 'superstar' woman scientist ends up on the desk of an employer, she may very well be hired; but we must pay attention to the signs that such high standards are not as much being demanded from men. Perpetuating such double standards leads to us losing out on the women who are capable but lack such rock-solid credentials. Such as the woman who was held back without any support and a sick child to raise; the one who lost the chance at a prestigious PhD because a sexist member on the male-dominated hiring panel assumed that she would drop out after marriage; the one who lost time recuperating from a stress-induced mental breakdown; the one who had to take a break to support her partner in scaling new heights; or the one whose physical disability stopped her from publishing as often as her able-bodied peers. Who is hiring these scientists?

14

What about the Kids?

Having been largely male-dominated for many years, scientific workplaces in India settled into the comfortable tradition where the domestic needs of scientists were taken care of by their wives. In a biography of C. V. Raman's wife Lokasundari Ammal, written by his grand-niece Uma Parmeswaran, we see a literal example of how much historic men in science have relied on their wives for support. The biography describes an incident where Raman was upset about going to work during floods in Calcutta. Seeing him helpless, the story goes that Lokasundari 'grabs two stools and makes Raman walk on the improvised bridge, herself wading through the water, changing the position of the stools until he reaches high ground'.[1]

Centuries of women's emancipation movements haven't changed the mindset that places the responsibility for unpaid care work on women—it doesn't matter if she is a working professional or not. This is the case worldwide but studies have shown that the burden is greater than usual for Indian women. According to a 2015 report from McKinsey Global Institute, while women all over the world spend triple the amount of time on unpaid work as men, in India, women spend ten times as much as men on unpaid care work.[2] Data from the Organisation for Economic Co-operation and Development (OECD) reveals that Indian women spend approximately 352 minutes a day on unpaid work—including elderly and child care—as compared to that 51.8 minutes by men.[3]

It is particularly disturbing when scientific workplaces, in which the nature of work tends to be especially unpredictable and time-consuming, function under the illusion that childcare is not a part of life. They got away with it for several decades because the mostly-male scientists had wives who were expected to provide them with unrequited support. This afforded the male scientists the luxury of single-minded commitment to their labs, which quickly became the gold standard. The idea of the eccentric and obsessive scientist with no time to engage with the mundane got glorified. Now, to the bewilderment of the still male-dominated science community, some wives and mothers are scientists themselves.

As more and more women enter the workspace, institutional structures have been forced to evolve. Maternity leave, institutional daycare and re-entry (after career break) fellowships were introduced to accommodate women. These are frequently brought up to exemplify how women-friendly science has become but there are large gaps in the design and implementation of these policies.

In 2008, Mayurika Lahiri returned from Harvard Medical School to start off her scientific career at the new IISER in Pune, only to discover that there was no daycare facility for her young daughter. The young scientist was stumped. How was she supposed to put in long hours in the lab if her child had nowhere to be? Luckily, the sympathetic Director assigned her a room close to her lab so she could visit her baby during feeding times. Though things worked out in this typically *jugaad* way that is so admired in our country, Mayurika did not forget about the near-crisis she was in. 'Everyone wanted a daycare on campus, and our director was very open to it, but someone had to take responsibility for doing this,' she recalled. Mayurika raised her hand even though by then her daughter no longer needed one; she led the effort to set up an in-house daycare centre at IISER Pune. The centre was inaugurated in 2015 and became immediately popular.

The setting up of a daycare centre at IISER Pune had a domino effect. This motivated other institutes to finally comply with the Maternity Benefits Act that mandates the setting up of a creche facility in every establishment employing 50 or more people.[4] After we published a feature on Mayurika's 'daycare revolution' on thelifeofscience.com, Mayurika was approached by IIT Delhi and several other institutes for advice on setting up their own creches.

> I set it up for the general good, for other faculty and staff. When I'm working, I don't even think about my child because I know she's safe. Our research benefits from the daycare centre. I am able to spend more time here . . . till 6.30 p.m. the whole time is for me and my students.[5]

The relief of having access to an in-house creche at his workplace was evident on Bittu K's face when we caught up with him for a quick online interview in 2022. The crèche in Ashoka University was a result of a couple of years of planning, and for Bittu, a neuroscientist and new parent, it came right on time. The challenges of parenthood hit those in non-traditional family structures especially hard. 'For queer and trans folk, it's a struggle to even be allowed to be a parent, because we don't have the same adoptive rights,' pointed out Bittu. If they are able to navigate through the social and regulatory hoops to adopt a child, childcare is only the next hurdle. A support system is not something that exists for trans people by default. 'Most other scientists I know, even women, have a partner and in-laws to help with childcare. We had to create a community of friends and chosen family to do childcare,' he added.

Despite the Maternity Benefits Act mandating creches in workplaces, institutional daycare is still not commonplace in universities and research centres. It often falls on the shoulders of parents in science themselves to set things up. In 2019, astronomer Preeti Kharb was tied up with the organization of the annual meeting of the ASI which was to take place in Bengaluru that year. Preeti, being an active advocate for gender equity in her field and a founding member of the ASI's Working Group on Gender Equity (WGGE), was particularly determined to make sure there would be childcare available to participants who needed it. As the day crept closer, we received a jubilant email from her informing us that the deed was done—'Childcare has been arranged by ASI. Such an achievement for us!'

Preeti's feeling of jubilation was justified because the quest for childcare had not always been an easy one. She and others had first brought up the matter of childcare at the general body meeting before the 2017 meeting at Jaipur. At that time, there was resistance from the senior members of ASI, and this included women members. 'Most of the members were also simply surprised that we were asking for this. This issue had never occurred to them previously. We got the impression that they had simply assumed that mothers and grandparents would take care of small children anyway. Their main concern was that nobody would leave their children in a childcare facility for safety reasons!' Preeti said. Subsequently, there was no childcare at the Jaipur meeting.

The following year, Preeti persisted. The meeting was to be held in Hyderabad and this time, fortunately, the main organizer was already sympathetic to the cause. However, there was no accessible daycare available at the venue, Osmania University, and private childcare facilities were too distant and expensive. 'Some of the participants who wanted childcare had to negotiate a decent price with these facilities on their own,' Preeti said. After the hiccups of the past two years, the relatively smooth journey in 2019 came as a big relief to her and the participants of the conference.

'At the previous meeting, I had already brought up the matter of childcare to the main organizer, who is a friend and a colleague. And he had promised to do this—presumably, because they had access to the necessary infrastructure, being in Bengaluru. It also helped that this colleague has small children of his own and a wife who is a faculty member at the university. In the end, persistence by the WGGE members, a changing world, and sympathetic members of the astronomy community were all needed to make childcare accessible at ASI meetings.'

The reason we know of Mayurika's and Preeti's stories today is that they had happy endings. Going by the number of institutes that continue to have no daycare arrangement for their employees, it seems very likely that many more such attempts to institutionalize daycare did not bear fruit, or perhaps the negotiations continue as we write. In the meanwhile, a vast number of young mother scientists in science have to rely on informal adjustments to do science and engage in scientific activities such as conferences that are often away from home. We witnessed one such adjustment and its impact in December

2018, during a national meeting for cell, developmental and reproductive biologists at IIT Kanpur.

The Indian Society of Developmental Biologists has a rich women-in-science history, having been co-founded by one of the pioneers of embryology Leela Mulherkar. Since then, it has been led by highly respected scientists such as the late Veronica Rodrigues and Jyotsna Dhawan. Such meetings are opportunities for the finest and emerging experts in the field to meet, network and exchange ideas. For the youngsters in the field who are fighting the highly competitive race for a place in academia, such meet-ups are of tremendous importance. Deepak Modi, a Mumbai-based reproductive biologist, was well aware of how crucial such opportunities would be for his team of grad students. So when one of them conveyed to him that she feared she would not be able to make it because of her six-month-old baby, Deepak decided to intervene.

'Usually, PhD students are given shared accommodations, but this would not do for Varsha*,' he said. Deepak got in touch with the organizers at IIT Kanpur and requested them to help.

'The organizer said that free accommodation would be provided for Varsha as well as a caregiver, and it was promised that this would be close to the venue so that she could easily slip out for breastfeeding. He also said that the on-campus crèche would be made available to her. In the end, Varsha made it with her baby and her mother!'

Though things worked out really well for Varsha, Deepak lamented that there was generally not enough support lent to women students, even in a field like reproductive biology. By and large, women PhD students are discouraged from getting married, he said, pointing out an example of a student who had actually been turned away from her previous institute for this reason. He admitted to taking some years to realize the responsibilities that come with being a mentor to a consistently woman-dominated team. Today, he has only praises to sing of his experiences with women teammates—mothers or not. To him, the Kanpur experience was clear evidence that often, when the structures are not already in place, it is just a matter of asking for support. 'Conferences are important for students to build their network. If you don't allow women to attend conferences, you are throwing her off the competition,' he said.

India is in an interesting juxtaposition where in 2022 more and more women are being encouraged to have professional jobs, but taking care of the family continues to be as sacred a duty as ever. As a result, 'supermoms' who manage it all are exemplified. There is one figure that receives as much, if not more, admiration than the supermom in science—the scientist–father. While it's pretty clear that parenthood doesn't visibly interfere in men's scientific lives as much as it does in that of women, there are, of course, exceptions.

That's what first drew us to Sujay Ashok, a string theorist at the IMSc in Chennai. We heard that he homeschools his kids at the institute and were excited about the prospect of a story that would shatter some stereotypes about gender roles. This could be that spark of 'positivity' that some would argue this book sorely needs. While Sujay's story was heartening in a way, it also ended up being a poignant account of the ups and downs in the scientific life of his wife, Eleonora Dell'Aquila.

'I'm not sure how to describe myself . . . I feel like I'm cheating if I call myself a scientist.' An Italian citizen, Eleonora moved to Chennai with her collaborator and husband Sujay in 2011. Eleonora and Sujay met as students in the US and have since published many papers together. There was one crucial difference between them—Sujay has a permanent job at the IMSc while Eleonora, when we spoke in 2019, had practically given up on the possibility of a stable scientific career. She was scrambling to make the most of any avenue available to her, usually short-term fellowships. Eleonora was in the middle of one such stint in Turin, Italy, when we caught up with the couple.

For almost four years after their first child was born, Eleonora and Sujay lived separately. She was in Canada for her postdoc and he was in Chennai where he had just begun his new job at the IMSc. Though they met periodically, the separation was tough on all of them. 'It was hard to find childcare for babies below one in Canada,' she said. Though she cherished the freedom of flexible timings at her workplace, Eleonora often sensed that she was being observed. 'They would check what time I came and went . . . it was to make sure I was coping, they said.' But it irked her that the men were never monitored in this way. The double standards of her workplace were further exposed when she brought her son to work one day. 'He cried once, and I found out that someone raised a complaint about this. The Director told me there was no point bringing the baby to work.' Eleonora wondered what made the cry of a child more intolerable to her colleagues than the routine screams and yells from the foosball table.

'I'd even seen that when men visited from the outside with their kids who roamed around freely, there would be comments about him being a 'cool guy' but to me, they said there cannot be 'unattended children in the workplace'. When I countered them saying he was not unattended, they said 'then you are not doing your job'.'

After their second child was born, Eleonora began making plans to move to India to reunite the family. 'I was depressed and it was getting too hard on the kids.' In Chennai, she managed to earn a couple of postdoctoral and teaching positions, after which she took some time out to have their third child. The Indian experience has been bittersweet for Eleonora. While she had not been able to procure a full-time position due to certain legal technicalities, she was heartened to see that scientist–mothers were given more respect than she was used to in the West. 'At first, it really did seem that India had it better. Women with children, in fact, are treated better than women without children!' she chuckled.

Though the Indian government mandates crèches in all establishments, compliance is extremely poor. For the first few years, Sujay and Eleonora sent their kids to a daycare centre in a nearby locality. Years later, when the couple decided to homeschool their kids, the IMSc was remarkably supportive of it. 'There was still no crèche in IMSc so we requested a room, far away from workspaces, where we could take turns to care for the kids and keep some books,' she recalled. However, this didn't happen so their kids hung around in Sujay's cabin itself. The arrangement worked out with far fewer complications than one would expect; in fact, things were going so well that Sujay would get the occasional request from other scientists to have their kids join as well! 'I'm not sure this would have been possible in Italy,' admitted Eleonora.

The story of Eleonora and Sujay brings out the many, often conflicting, nuances of childcare in Indian academia. On one hand, most research institutes in India continue to have no in-house daycare system. However, the sanctity of the family as a unit has made colleagues and bosses empathetic towards scientists with childcare needs. This informal support is especially useful in the absence of infrastructural support. 'Most people cooperated in my case,' said Sujay. 'But it may have been difficult if it was a woman scientist doing this,' he conceded, remembering what his wife went through in Canada. Nodding,

Eleonora added, 'In Canada, there would be more structures by law though. In principle, I feel I could have written to someone that I'm being discriminated against. In India, on the other hand, there are grandmothers, maids and cooks.'

The success of informal set-ups like Sujay's suggests that the assumption that childcare makes scientists slack off at work is wrong. Provided with the right support at the right time, scientists with young children have for decades been doing science with no problems. Smitha Hegde, a Mangaluru-based pteridologist, told us that the young mothers in her team are much more 'aggressive' than the rest.

'I think this is because they are more determined to make something out of themselves. Babies make them stronger. They have more focus. As long as someone will take care of their baby, that is—either mother or mother-in-law or a nanny—that is a must. I have had a very good experience with pregnant women and mothers. They have a deadline and they know what's coming so they plan well and execute it beautifully.'

Perhaps the reason Smitha is so supportive of scientist–mothers in her team is that she herself was once a beneficiary of a similarly progressive mentor. When she started her career in 1991 as a lab assistant at Mangalore's Aloysius College, she had a baby to take care of.

'There was a very noble person in my building, Rev. Fr Dr Leo D'Souza. He really supported women and girl students. You know, I could get my baby to the lab! He made a bed in a corner of the lab where my baby could play, sleep and be fed. I owe him for supporting me.'

It's bad enough when institutions and leaders uphold and defend patriarchal family values that thrust upon women the primary caregiving role but to expect women to perform the same as men who are free from these responsibilities does not make sense. Even less so when the systemic support is not there. The few women scientists who have overcome this unfair competition are usually those blessed with an exceptional degree of informal support. Often in the forms of 'grandmothers, maids and cooks', as Eleonora astutely observed.

Every so often there is a sign that policymakers are aware of the gaps that exist. We hear of a new policy or charter or scheme to address these. But even so-called 'women-friendly' policies (as if childcare is something that the father need not bother with) seem to have been framed in a spirit of 'this will help *them*' rather than 'this will help *us*'—implying that academia still belongs to men and childcare is still the responsibility of women. Bittu K. believes that changing the language of childcare policies can make a difference in perceptions, and make life a little easier for single and queer–trans parents. Enabling and encouraging paternity leave alongside maternity leave can signal that fathers are expected to contribute equally to childcare, he said. As a transgender person, Bittu is grateful that his institute had a provision for a gender-neutral 'parental leave', and this was available to adoptive parents like him as well. 'The standard academic structure assumes that 'the scientist' is not going to do childcare,' said Bittu, adding, 'So you see male scientists saying yes to an 8 p.m. meeting and then it's up to the women scientists—plus me—to say this is not acceptable. Now some of us have come together to say you can only expect us to be present during creche hours.'

15

The Old Boys' Club

The sun hangs low and dim at 11 a.m., typical of a January morning in the north Indian plains. We are trying out the bread pakoras at the small IIT Kanpur canteen that shares a wall with the less popular Cafe Coffee Day. The *chhota* canteen looks over a sprawling amphitheatre surrounded by academic buildings. Groups of students are sitting around chatting between lectures and experiments.

As we look for chairs to park ourselves with our steaming fried food and tiny cups of chai, a few people covered in mufflers are also shuffling for chairs. They pick some up and take them over a few metres to join a large seated circle, adjacent to the amphitheatre. The circle is composed only of older men, presumably professors and staffers of IIT Kanpur. Not a woman in sight! Casual banter and giggles are passed around the circle; every few moments, roars of laughter erupt from different places in the growing circle as new entries join in. This is a common site at IIT campuses. It is presumably a physical manifestation of the old boys' club that many of our interviewees have talked about. To our 'female gaze', if you will, the sight of the boys' club invokes some measure of amusement as we watch as outliers.

None among the hundreds of scientists we have interviewed in the last six years would deny the existence of the old boys' club in Indian science. Misadventures of women in science with the club can include helplessly watching their harassers enjoy impunity. It can also include simply not getting a science job they are perfect for. According to a reader who commented on our website, 'gender bias of advisory committees' prevented her from securing a job in the country. The domination of men in science can also mean long spells of isolation like the physicist we met, who being the only woman in her department since she joined in 2010, felt consistently 'vulnerable among all the men' and 'overlooked.' And at the other end of the spectrum, we heard admissions like the one voiced by Rashna Bhandari from the Centre for DNA Fingerprinting and Diagnostics (CDFD), Hyderabad: 'I have not seen the boys' club in my close academic circle but I know it is there.'

The boys' club is simplistically understood as a social phenomenon—something humans do. Everyone likes to stick to their kind. No one wants to feel like an imposter, especially when your career is on the line. A large study that analysed the gender of co-authors of 9.15 million scientific papers indexed on PubMed reported a 'gender homophily' between scientific collaborators. It reads:

> Evidence suggests that women in academia are hindered by conscious and unconscious biases, and often feel excluded from formal and informal opportunities for research collaboration.[1]
> Our results reaffirm that researchers co-publish with colleagues of the same gender more often than expected by chance, and show that this 'gender homophily' is slightly stronger today than it was 10 years ago.

So men stick with men and women stick with women; for some, it is even a necessity: trans-gendered folk must group together to assert their mere existence. There is groupism in science, as everywhere, but the trouble is groups that make decisions affecting everyone are composed mainly of cis men.

Saman Habib, a malaria researcher, recounted her experience with gender-based segregation when she first joined the Central Drug Research Institute (CDRI) in Lucknow in 1994. 'I chose CDRI because of its historical significance as a public health institute. It had been built in 1951, as the newly independent nation prioritized work on infectious diseases rather than looking at other things.' In 1998 at CDRI, Saman shared in an interview, there were a good number of women, at least for those times.

'There were quite a few women. But there was a clear binary of behaviour between the genders. At gatherings, say an annual day or a retirement party, the women were huddled on one side and the men on the other.'

'Also, the women were not vocal in meetings. And I could see from the faces of these really old conventional senior men in the hierarchy, that they did not like them being vocal. Even if women were polite and spoke in a sober, straightforward way, they did not like women expressing opinions,' she added. For Saman and her husband Amit who had also secured a job at CDRI at the same time, this gender-based groupism was alien. They had just wrapped up their PhDs at the National Institute of Immunology (NII)—a fairly new institute at the time, the first one that the Department of Biotechnology (DBT) took under its wing. 'NII was a very vibrant place, very modern in its outlook for that time. It was the first place in India to have a common hostel for boys and girls,' she said proudly. 'Much was made of the fact that you were in this premier institution, and that you had independence as PhD students.' After NII, the more conservative CDRI was exceedingly frustrating for the couple. So, at official gatherings, Saman and Amit started to resist gender-based segregation.

'We would cross over the gender barriers because we were just so irritated with this. He would come to the women's side and I would go to the men's side. And we sort of started to break the mould.'

'And after so many years, now I am so happy to see that such segregation just doesn't happen anymore. Because times have changed, the culture has changed and, you know, things are so different, but they were not always this way.'

Segregation based on gender might be a thing of the past at some institutes but it is still far too common at others. As Saman herself has experienced, the arrow of time doesn't always bring changes uniformly across institutions and even when it does, it doesn't mean they are here to stay.

There is a long list of men with ample Western connections credited with setting up modern Indian science as institution builders. J. R. D. Tata started the IISc in Bengaluru, Homi Bhabha set up the TIFR, and Bhabha Atomic Research Centre in Mumbai, C. V. Raman founded the RRI in Bengaluru, Vikram Sarabhai had a big hand in setting up what is now ISRO and J. C. Bose, the founder of the Bose Institute in Kolkata, are some among many others. These gentlemen amassed capital in colonial times and mobilized it into the institutions where our top scientists work today. Around this time, women were only negotiating their enrollment into these institutions as students. After them, came the second generation of science leaders taking forward the expansion of India's reach in global science. These were also men. Women have been very low in numbers and were hardly ever let into the leadership leagues. This reality is apparent even today from the current list of directors, vice-chancellors and government secretaries in science ministries. Institution building in Indian science continues to be male-dominated.

'Academic pedigree is very important'. This is a common line we heard repeatedly whilst lab-hopping. Often, it is offered as well-meaning advice to young graduates starting out in science. What this statement implies is that one has a lower chance at great success if one isn't in any way connected to the institution builders of Indian Science. Those few men. A modelling study published in *Nature Communications* in 2019 provides the statistical proof behind this trend. Its self-explanatory title reads: 'Early Coauthorship with Top Scientists Predicts Success in Academic Careers.'[2]

The authors discuss their finding that early co-authorship with top scientists can have a long-lasting effect in this way:

> It is tempting to relate the above results to the 'Newton hypothesis', i.e. the idea that Science mostly progresses thanks to the work of a few elite contributors who typically 'stand on the shoulders of giants', i.e. who rely on the previous work of other elite scientists (as opposed to the 'Ortega hypothesis', which instead purports that scientific progress mostly comes from the incremental contributions of many average scientists). A number of studies on citation networks provide empirical support to the Newton hypothesis, showing that newly published papers tend to lean on a handful of important past contributions in their field, and that disciplines are organized around 'rich clubs' of top scientists that tend to preferentially cite their peers' work. Our results also go in that direction, showing that scientific elites play an exceedingly important role in shaping the academic landscape at all its levels.

'There's a very, very strong old boys network,' Gagandeep Kang, ex-Director of THSTI Faridabad also tells us. 'Just today, they wouldn't hear me,' she said when we caught up with her in 2019 while she was still heading the institute. 'I was chairing a meeting today. And the men were talking over me. Some older men seem to think no matter what their role on a committee, they are the authority figures under every roof.'

Calling out the old boys' club as the biggest instrument of 'institutional' sexism operating in our research hubs has been done so explicitly by only a few women like Gagandeep Kang.

One way to say the same thing in a presumably secure way is: 'In any research council meeting or committee meeting you will meet the same people.' This line was oft-repeated during our interviews. It is common knowledge that the old boys' club controls access to the small fraction of public money and infrastructure that has been set aside for India's scientific research. The men on top and their networks not only dictate who gets to do science in India but also what research Indians pay for. Gagandeep elaborated: 'To get funding and other opportunities in the science world a lot of the decision rests on whether or not the review panel consists of people who know you. It's a lot of who-you-know and how-you-know-them.'

First, she noted, there is always a long delay in you being put on committees if you are from outside this circle of influence of the boys' club. Then, the opportunities one needs to get ahead in science are hard to come by. To get to the illustrious positions at the top of Indian science she occupies today Gagandeep, who started as an outsider herself, has made it through quite an obstacle course. She believes it is not difficult to quantify the opportunity bias because of gender-based discrimination in one's own career. She has done so for herself and comes up with a figure of a five- to six-year delay in reaching where she is today.

'If you compare the support that is given to faculty members at early stages in their career in terms of travel opportunities, lab space and consumables, you can see that women usually get just two-thirds or less in every sphere than men in their career. This is true when it comes to opportunities to build your scientific network as well.'

'It's who you take to meetings with you. Men are more likely to take younger men than women to professional meetings that would result in exposure and advancement. So it takes women longer to build up networks. And then there are opportunities for sitting on committees, for being part of a team that sets up guidelines and frameworks that give you standing in the field. The number of women being put forward to participate in those is very few. It is required for women to be at a much higher level to begin participating in decision-making positions,' Gagandeep said.

In 1992, when she was starting a research project to characterize various pathogens, Polymerase Chain Reaction or PCR was being established in labs across the world. PCR is the same nucleic acid amplifying biotechnology that COVID-19 has made familiar. 'The technique was just beginning to be used in biotechnology. There was almost nobody in India, I think, who had a PCR machine.' At CMC Vellore where she did her PhD, there came an opportunity to get trained in running PCR tests. Also eligible to travel abroad for this training, was a male colleague—'this person doing work in the chemical biology space.'

'Now to the question—who is more relevant for the opportunity? Very clearly, it should have been me. Instead of which, 'our man' gets to go for the training program and never uses PCR after that.'

'Okay . . .' she continued... 'I finally learned PCR six years later, after I went to London for a postdoc. And if I had learned it before, I could have used it for my previous work. That investment in me would have yielded much greater returns. I could have gotten to where I am now much earlier.'

'Yeah. That didn't happen,' she summed up.

The old boys' club is entrenched deep into the scientific system and is not restricted to the top institutions. A scientist, on the condition of anonymity, told us of a horrific incident that illustrates how deep the influence of the boys' club can go. A husband–wife scientist couple working in the same institute had ended their relationship bitterly with the wife emerging out of the conflict with wounds on her body. After being violently attacked by her husband–colleague, she sought to join a new position at another institute in a different city so that she may move away from the abusive environment. The Director of the institute signed the transfer documents but attempts were made to stall the paperwork at other higher offices of CSIR— the governing body of both institutes—in order to pressure her to withdraw her legal complaint. This naturally caused much despair for the victim.

Another example of the extending arms of the boys' club we came across in an interview with Ramadevi Nimmanapalli. When she returned from the US in 2014, she was tasked with starting the veterinary faculty at BHU. By joining the university she raised the count of women directors/deans in BHU from one to two in 16 faculties. She had her work cut out for her. As she was brought on board to build the veterinary faculty and revive the attached dairy farm, there was work to be done and she felt confident. The university dairy with over 600 cows was functional but suffering losses, and there were hardly any experts to take care of the animals. New to the way things work in Indian science, she got on with her to-do list, which involved various clearances with the UGC and the Veterinary Council of India, with the straightforwardness she had learnt while working in the US. Whenever required, she walked into the VC's office with prior booked appointments and asked him for the signatures she needed. Among the buildings Ramadevi built was a veterinary hospital adjoining the dairy farm, which supplied the various canteens at BHU with milk products.

'I put in all the facilities needed where doctors would be comfortable and purposeful in maintaining the health of our animals—hygienic tiled walls, for example—anything that was required to make a functional department. There was also scope for research in collaboration with the doctors,' she told us during an interview in her office in 2017.

But soon things changed. When a new VC took over the leadership, Ramadevi's streak of accomplishments took an unexpected hit. 'There were many who worked under the previous VC who viewed me as an outsider . . . an outsider woman.' Some BHU staffers perceived Ramadevi's swift progress as odd, a sign of something fishy that they found unacceptable. With the change of power at the top level, they took the opportunity to attack Ramadevi. Letters were sent to her office accusing her of having an affair with the previous VC. The animal hospital she set up was forcibly shut down and locks were placed on all access doors. It was declared defunct for no apparent reason or explanation by people Ramadevi chooses not to name.

'Ultimately, the new VC gave the authority of the farm to someone else. The hospital is still being used as a godown for cattle feed,' she said. Immense public resources were wasted and rumours were kept afloat only to challenge her power.

Ramadevi herself succeeded in establishing the faculty of veterinary and animal science at BHU with 17 departments and 67 faculty members housed in modern buildings that promise to be the hub of cutting-edge veterinary research in the country.

Ramadevi had moved on from her role at BHU when we caught up with her in 2022. In her reflections on her time there, we could read a sense of pride for what she had achieved along with a feeling of relief of having moved away from the experience. But her spirit is not broken, as she suggested she is up for another challenge for India if and when she gets an opportunity.

A retired biologist from a premier institute puts the above cases in perspective better than we could ever hope to.

'Science in India also has become a power game. It is not assessed by scientific temperament or ability. Scientists are not allowed independence. If you aspire to do science and have ideas of your own there is active suppression and the ones not in the power game are treated like technicians, as if you have been recruited to work for the Institute's 'group leaders'. The funding allotted by the government of India is shared among a few in the club and others are left high and dry. It is like 'water water everywhere but no drop to drink' type of situation. *Jiski Laathi Uski Bhains.* (Hindi idiom: Who holds the staff, owns the bull),' she summed up.

Institutional sexism plays out majorly as an opportunity for bias in the Indian science community. When any opportunity lands on the desk of someone up the ladder, it is forwarded into the boys' network. Scientific awards are the clearest example of this.

Indian science awards and prestigious fellowships are notorious for inexplicable age limits as well as a very low number of women awardees. The Shanti Swarup Bhatnagar

(SSB) Prize is considered the top scientific honour in the country as it is a national recognition awarded by the biggest scientific establishment in India, the CSIR. Along with the esteemed recognition, the award offers five lakh rupees and a subsequent lifetime of stipends. As of 2021, only 19 women had ever won the SSB prize in its 63-year history. No women won the award in 2021. The Swarnajayanti fellowship, another top honour, this one awarded by the DST since 1997, is given out to around 15 scientists in different disciplines every year. In the past decade, the number of women who won the fellowship flickered between zero, one and two with no sign of a trend that might indicate the situation is getting better. Even the supposedly progressive Infosys Prize has a disappointing gender ratio, in 2021 no women won any of the Infosys science prizes in the four natural science categories.[3]

Behind every scientific recognition, be it an award or a prestigious fellowship, is a large number of nominations, nominators and jury, much like the Nobel, Oscar or Filmfare awards. For example, to compete for the prestigious SSB Prize, first, one needs to be younger than 45 years of age. Then one must secure a nomination from either a past award winner or by an official of some position—Secretaries in the Government, Presidents of the academies, VCs and Directors of institutes etc.,—positions that are no doubt dominated by men. Only then can you begin working towards the large bundle of paperwork with all your publications, achievements, and what-have-you to prove your worthiness for the award. The convinced nominator of repute must then sign off this bundle towards the CSIR-based (secret) award committee that will choose the winners. The identities of the members of the SSB Prize committee at CSIR are kept hidden; it is said to avoid untoward schmoozing by prospective awardees.

In the case of the prestigious Swarnajayanti fellowship, self-nomination or sending an application is permitted. If you win, a boost to your salary, along with a five-year grant of Rs 5 lakh each year plus other perks is promised by the DST. But to win you must be less than 40 years of age and convince two rounds of interviewers—first, a subject committee composed of people of repute in your own field of research and then a national committee of top scientists making the final decision. Even in the case of the Infosys Science Prize, or India's Nobels as the founders would like it to be seen, all the jury chairs have until 2022 been men. Though it has a more progressive veneer and a more international appeal, here too, the identities of nominators stay hidden behind the words 'invited nominators'. The absence of women among the jury chairs and board of trustees is made all the more obvious by the larger-than-life posters and stage formalities at the lavish award ceremonies held year after year in Bengaluru. Indeed, as we went looking for Indian women in science, such realities made us cringe.

In 2018, Aditi Sen De became the first woman to ever win the SSB Prize in Physics. At the Harish-Chandra Research Institute (HRI) in Allahabad where Aditi works, you will find her in the institution's pantry more often than in her office. Every morning after her daughter is dispatched to school, she proceeds to the HRI pantry with her laptop. Her students start turning up there around noon and all day the QIC group—Quantum Information and Computation discussions happen here. Often, they last till midnight. 'I have an office, but I hardly use it. In the pantry, I can enjoy our nice and peaceful campus as well as get chai and snacks at any time,' she said in an interview. When we brought up her SSB Prize, her first reaction was of extreme embarrassment.

'I feel embarrassed because I know so many women who worked before me and my contemporaries that also deserve this prize,' she said. Although Aditi was truly earnest, this statement from her means a lot more than what one might say while gracefully accepting an award. 'I really hate that no other woman physicist has the recognition and encouragement I have now. That's why I feel embarrassed. There are so many of them that could have got it. But they didn't.'

'Sometimes, I feel these awards are like tossing a coin and I got it in a similar fashion,' she added.

As a winner of the SSB Prize, Aditi now has the power to nominate someone else to win the award. In the months after receiving the award, she became aware of how crucial her status is as the first Indian woman physicist to ever receive this celebrated award. She said, 'I wasn't aware of this but now I understand that since there are hardly any women (in the club), women are not able to nominate women. Only if the men think that a woman is good then they will be nominated. Actually, only a few women manage to convince these people.'

At a physics meeting where we met her first, Aditi in her post-award-winning glow delivered a popular lecture on QIC. The resounding applause at the end of the lecture hailed not only her scientific work but also her remarkable feat of breaking through the boys' club barrier by winning an SSB Prize. Over dinner that day, Aditi expressed her concern for the 'growing' younger women in science whose work might not be recognized in time because people who are on committees today don't know them or don't like them. Many women scientists do not go on to become established because of this, she believes.

'Some people who are 'famous' in a particular field, have a bit of power over outsiders. I stand no chance if these individuals say to others in selection committees that my work is not that important, not worth the award or "it is good but it is not all that". Someone could question this judgement and point out my publication record. To this, the response from these famous people often is: "She is publishing but the work doesn't have any impact." Then a committee member not in my area of expertise is unlikely to go read all my publications and decide for themselves. not everybody has time for that.'

And then that judgement sticks . . . until . . . if you're lucky it doesn't, like Aditi.

Aditi faced such a dismissal of her work herself in the early years after she returned from a lab abroad in 2009 and set up a QIC group at HRI.

'When I joined, there were very few QIC conferences going on in India at the time and many people, even contemporaries my age, my juniors and seniors, were invited. And I was working in the field too, had publications, a position and students, but I was just excluded from those conferences. When this happens you feel bad and spend a lot of time thinking about what is wrong with yourself. You feel quite disheartened.'

In time, Aditi was able to overcome self-doubt and see that such an exclusion was purposeful by certain men in the old boys' club she will not name. 'But we have to keep in mind that not everyone is equipped to surpass all doubts, take the gender bias in their stride and continue working,' she said.

Things are no different at the other flagship scientific recognition by the Indian science administration, the Swarnajayanti fellowship. 'When it comes to national-level fellowships or awards, I often find myself facing all-male committees where there is a high possibility of a bias, be it conscious or unconscious,' another scientist shared her frustration over the old boys' club. Heena* just turned 40, now making her ineligible for the Swarnajayanti fellowship. Recently, she sent in an impressive application that came with the necessary research proposal for a five-year project to be conducted at her established lab at a premier Indian institute. It took her straight into the first round of interviews with the subject committee. 'These were seniors from my field and, off the cuff, I was told at the end of the interview that I was at the top of their list.'

'At this interview with the subject committee, there wasn't any particular question related to gender as such but that wasn't the case with the next one, which had a different body language,' she said during an interview in her office in Mumbai. The final round of interviews with the national committee was juried by those few people 'you always see at top committees in Indian science,' she mentioned.

'It's an all-male committee. And so, of course, there were a lot of doubts and discussions on the fact that I have another project running. Actually, I have four, as all large labs do multiple projects at one time and they are all connected—it is not as if I am doing cancer biology in one and archaeology in another.'

Heena* did not win the fellowship. She didn't even feature in the list of top two in her field in the end. 'Two others got the fellowship, both of course from the male gender. I wasn't even considered good enough to be in the top two, while I was actually number one, to begin with, from the discipline's perspective. I wouldn't say it's completely because of my proposal, because that was vetted by the people who are known as being the top-ranked in my field. So then what else can be happening?' she wondered.

'From the kind of discussion I had with the all-male national committee, I understand that they thought I was already involved in a major project and shouldn't be taking on more work. So the question is how much can a woman do?' she added sourly.

'Ultimately, little can counter good science,' Heena goes on to believe. She has received several top recognitions, of which none are from Indian science establishments.

We've heard several women, especially those who are internationally recognized, express their disappointment over reserving recognition for women scientists to the Women Scientist Awards. The converse effect of there being a women scientist award is a general perception that the so-called gender-neutral awards are only for cis men since there's a separate award for women. Rashna Bhandari, the molecular biologist who had claimed she has not yet encountered the boys' club personally, said: 'DBT in all its wisdom, has instituted two prizes, the National Women Bioscientist Award and the National Bioscience Award. One of my male contemporaries, giving me very well-intentioned advice, said 'Hey Rashna, you should apply to the Women Bioscientist Award'. He failed to suggest that I should apply for the National Bioscience Award. He may have assumed I am eligible for or stand a chance to win only the Women Bioscientist Award.'

What's worse is when Heena revealed to us that she had in fact received a call from the administrators of the DST's Young Scientists Award after she applied for the same, explaining that they would like to move all women applicants to the Women Scientist Award category!

Hearing from women scientists who are seeking recognition and are turned away, we realize that 'women in science' issues usually only bring up topics like attrition and measures needed to 'help' them. Forgetting to see the excellence in the work of women scientists while limiting women in STEM discourse to the many problems of women exposes the benevolent sexism prevalent in the Indian science community. Are recognition and excellence reserved only for men? Many women scientists we met are asking.

'Somebody might remark that 'we are harking back to recognitions—it's not why we do our science, is it?' This might be true, but if we aspire for prestigious recognition, what is the problem with that? Should we only just stay happy about being allowed to do science?' Heena asks.

On 26 September 2022, the scientific community eagerly awaited the announcement of the year's SSB Prizes. The day went by with no winners announced. Instead came a shocking announcement that the majority of the awards given by various ministries to researchers in the country were to be scrapped. The SSB Prize, according to reports, had been spared, though the year went by with none being awarded. There was no clear explanation of the rationale behind this sudden decision to scrap over 300 awards, except that the intention was to restrict awards to 'really deserving candidates'. Officials also shared that they were in the process of instituting a 'Nobel Prize-like award' of 'very high stature' called Vigyan Ratna to replace the cancelled awards.[4]

16

A Taboo around Quotas

In September 2019, the University of Hyderabad played host to a special physics conference. Special because it featured the voices of not only subject experts but also sociologists, educators, administrators and bureaucrats. The three-day event was to be 'a first-of-its-kind national meeting in which both physics-related and sociology of physics-related topics would be deliberated'. Naturally, as science journalists interested in gender, we were there, eyes and ears open, pens perched on our notepads. During a panel discussion on women in science, a senior scientist on the panel dropped the phrase—'we don't want quotas'. The scientist was referring to the contentious idea of reserving seats for underrepresented groups. Their choice of words irked some members of the audience including mathematician-turned-educator Jayasree Subramanian, who was prompted to stand up and ask, 'Why can't we have reservations?' When we caught up with her later that day, she said,

'Again and again [during the conference] there was repeated reference that 'we will not have reservation, we will not have quota'. I think that is totally uncalled for. Who are we to decide? Reservation is one of the most important things that

India has done. We must talk about reservation and the politics of reservation.'

Jayasree is a vehement opponent of the idea that including more people from the margins will dilute the quality of science. On the contrary, she feels it will bring a much-needed transformation to Indian science.

As much as it feigns commitment to gender equality, the predominantly upper-caste Indian science community typically turns up its nose at reservation policies. Such measures are viewed as too extreme and problematic and a more 'organic' evolution into equality is preferred. This is a reflection of global science's attitude towards quotas. Following the 2021 Nobel Prizes, which recognized zero women, Göran Hansson, the then Secretary General of the Royal Swedish Academy of Sciences, said,

> We have decided we will not have quotas for gender or ethnicity. We want every laureate [to] be accepted . . . because they made the most important discovery, and not because of gender or ethnicity.

He added, 'We have discussed it . . . but then it would be, we fear, considered that those laureates got the prize because they are women, not because they are the best.'[1] Those in positions of authority argue that their resistance to quotas at any level— education, jobs or awards—reflects their commitment to meritocracy and to maintaining the quality of science. So embedded is this logic in the psyche of the academic community that being pro-reservation has come to carry with it a stigma. Despite reservation policies being backed by the Indian Constitution, most working scientists in India today wouldn't be caught supporting reservations and quota, not with a ten-foot pole.

In her memoir *Coming out as Dalit*, journalist Yashica Dutt makes a very compelling point: upper castes themselves have been benefitting from reservations for hundreds of years. Not in the form of quotas, but in the form of 'connections, networks and contacts'. She writes:

> Before international companies set up shop in India (and often even after) it was customary to pass down the father's job to the offspring, in both private and public sectors. But while most Dalits would pass down their lower-level office jobs to their children, upper-caste employees were handing over high-level managerial jobs, some of which would stay in the same family for generations . . . This upper-caste reservation, where most high-paying managerial jobs would pass on to the employees' children or relatives, was the way most companies hired candidates. Even today, 60–70 per cent of jobs in the private sector are secured because of family connections and networks.[2]

So then it must be asked where this upper-caste disgust for reservation policies is coming from.

While the word 'merit' is thrown about with abandon, very rarely, does the scientific community reflect on what it means? Merit in science is usually defined by a number of factors: for example high marks in exams, high-impact publications, strong recommendation letters, foreign education, conference participation, awards and so on. But do all individuals have an equal chance of fulfilling these criteria? Absolutely not. In the race for scientific glory, some groups undoubtedly enjoy an unfair advantage, one that is rooted in the historical oppression of other groups. The scientific community's refusal to acknowledge this ensures that they are able to separate 'merit' from underlying social realities. As a result, we have no qualms about accepting someone as a brilliant scientist even if he is an infamous sexist, a known casteist or a sexual predator. We are not discomfited by the star status of an institution full of award-winners, even if it has a miserable gender or caste representation. Encouraging sociopolitical consciousness among our scientists and students of science could be hugely beneficial, but doing so is a threat to the status quo. Very quickly, the broad gaps in India's self-proclaimed meritocratic science establishments would be exposed and this is something that the current wielders of power are afraid of.

We observed during our research that the more prestigious the institute, the more resistant the attitude towards reservation. It's not that these institutes are opposed to diversity, it's just that they believe that representation—whether gender or caste—will improve organically and not via supposedly merit-threatening artificial measures such as quotas. In 2017, something changed. The IITs, the country's elite engineering institutes, acknowledged that an organic evolution into a gender-equal space was, in fact, elusive. The IITs had gotten away with being heavily male-dominated since their founding in the 1950s with the stubborn belief that as more and more girls received primary education, gender ratios would improve. But even after six decades, this was not happening.

Every year about 10 lakh students attempt the Joint Entrance Examination (JEE) Main, the first qualifying examination for admission into the top engineering colleges in India. Of these, only the top performers are eligible for the JEE Advanced, which is prided as being one of the toughest entrance examinations in the world. The JEE Advanced is the qualifying exam to win a seat at an IIT and only 1 per cent of those who attempt it, make it. A 2012 report published by the Joint Admission Board

(JAB), the board that conducts these exams, stated after the JEE Advanced exam that year:

> While only about twice as many boys registered than girls, eight times more boys qualified, nine times more boys were counselled and got admission offers. The same trend is seen across various categories with minor variations. Ratios for OBC are even worse and those of SC and ST show marginal improvement. To add to this, girls have got lesser ranks [in the JEE Advanced exam] also.[3]

Indeed, year after year, only 11 to 12 per cent of the students who qualify for IITs would comprise girls (compared with a national average of about 28 per cent girls in all engineering colleges). The leak didn't end there. In 2017, though 12.5 per cent of those who qualified for the JEE Advanced were girls, only 8 per cent of those who eventually took up their offer were girls. This attrition is presumably because the rest did not get into the stream and location of their choice. A girl from Tamil Nadu for example may turn down her seat all the way at IIT Guwahati, or many girls may opt-out of streams such as mechanical engineering which are heavily male-dominated and carry a perception of being unsuitable for women. This provoked the governing IIT Council, in their 51st meeting on 28 April 2017, to take action. They announced that starting the next academic year, a new scheme recommended by a JAB sub-committee would be implemented to boost female enrolment in the BTech programmes from the current 8 per cent to 14 per cent in 2018–19, 17 per cent in 2019–20 and 20 per cent in 2020–21. They would do this by creating supernumerary seats to bridge the gap between the girls who qualified in the JEE

Advanced exam (4570 that year) and those who end up accepting the offer of admission (848 that year).[4]

The key aspect of this new scheme, for those who hold the JEE meritocracy dearly, is the word 'supernumerary'. While reservation provides opportunities to the marginalized at the cost of some seats for the general quota, here, in the push for 14 per cent girls, nobody would be losing seats. For example, if only 80 out of 1000 admitted students are female, then this needs to be raised to 140 to make it 14 per cent. The additional 60 seats would be added to the 1000, so the seats allotted to male students remain unaffected. It's worth noting that at the end of this exercise, the percentage of girls will still be short of the promised target—140 out of 1060 is 13.2 per cent. But let's leave that aside for a moment.

'This is not reservation and it would be misleading to call it reservation,' emphasized computer scientist Timothy Gonsalves who led the sub-committee, in an email interview.

'Reservation has connotations in the Indian context. It implies that a certain number of the sanctioned seats is earmarked for some category of candidates. If there are insufficient number of those candidates, the cut-off criteria are lowered to make more eligible. In the case of female supernumerary seats, there is no change in the criteria and the sanctioned seats are not touched.'

While speaking with a director of one of the IITs in 2018 about the gender ratio at the institute, he too insisted that this scheme was not a form of reservation. 'We made it supernumerary and the algorithm is quite fair. The difference now is that girls who might have otherwise only got into the metallurgy stream (which is not popular), can now get into electrical or computer science (courses that have a high demand).'

Just in case there's still room for confusion, in its annual document of 'Business Rules for Joint Seat Allocation', the

Joint Seat Allocation Authority added this slightly desperate sounding note: 'Important Note: Creation and allocation of supernumerary seats for female candidates is fundamentally different from reservation of seats for female candidates AND should NOT be mistaken for any kind of reservation for females.'

Got the message? This is NOT reservation.

In case it isn't obvious, the IITs are not fans of reservation. They were one of the loudest participants in the 2006 anti-reservation protests, along with AIIMS in New Delhi. The detractors were arguing that reservations would discourage merit and dilute the reputation of these institutes. The widespread protests were not successful, and the Supreme Court upheld the law for the provision of quotas for marginalized castes in higher education institutes. 15 years have passed but authorities of the IITs still do not seem to approve of reservation. Somehow, the history and reasoning behind the ratification of these policies in the Indian constitution are lost on them. They seem to be ignorant or have forgotten that centuries-long oppression by upper castes has put Dalit, Bahujan and Adivasi people on the firm backfoot and allowed for the gross overrepresentation of upper castes in fields such as STEM. The same brand of Brahmanical patriarchy has sidelined women too. Rather than allying themselves with those previous movements to boost gender diversity in educational institutes, the IITs and its sub-committees go to clownish lengths to position their scheme as something very different from reservation.

The JEE committee strongly refutes any suggestions that this supernumerary scheme to increase the number of girls at IITs will lower the standards of their student pool. The committee wrote:

The academic performance of the BTech batch is expected to increase . . . Only female candidates who have qualified in JEE (Advanced) are considered. They are in the top 2% of all students admitted to engineering in India. Evidence such as Board exam results indicates that they are as meritorious as their male counterparts.

The supernumerary scheme continued till 2021 when the percentage of girl students hit 20 per cent. But what happens after that? According to the JAB sub-committee (which comprised 40 per cent of women but 0 per cent of social scientists) report: 'After that, with organic growth in the percentage of females admitted through gender-neutral [pathways], female supernumerary seats will no longer be needed.' We asked Timothy Gonsalves, who led the committee, how he could expect organic growth to be achieved when it had eluded them for so long. He replied that the boost to 20 per cent (a number designed to be significant enough to 'catch the attention of masses') will alter the mindset of parents who obstruct their girls from joining a male bastion. 20 per cent is somehow expected to fuel a societal change and create enough role models to encourage more of the JEE Advanced qualifying girls to take up a seat at IIT. And to a certain extent, he feels he is being vindicated. In 2019, IIT Mandi, of which he was the founding Director, became the first IIT to have 20 per cent of girl students in their BTech courses. According to Timothy, signs of societal change are already emerging:

'I have noticed a marked increase in interest in IIT BTech from parents who've talked or emailed me. There have been positive changes on IIT campuses by having a more normal gender ratio in UG. I taught the entire 1st BTech batch

'Introduction to Computing' every year from 2011 to 2018. In 2017 and 2018, with 15% and 20% females, the classes were noticeably more responsive.'

While this move was a foolproof way to boost the number of girls joining IIT—the admission of girl students rose from 9.15 per cent in 2017 to 19.8 per cent in 2021—not everyone is convinced. After the 2017 move was announced, Dheeraj Sanghi, a computer scientist and former professor at IIT Kanpur, took to his blog to describe how this would work and what this means.

> I am amazed at the capability of our academic leadership to give spin to something as obvious as reservation. You have a minimum 14 per cent seats for women in every program in every IIT and in every category, and yet, this is not to be called reservation, simply because in case of SC/ST/OBC reservation, if a candidate can get both an unreserved seat and a reserved seat of the same program, then s/he is given an unreserved seat, and in case of women, she will be given a reserved seat.[5]

Dheeraj was referring to how in the case of caste-based reservation, the candidate can also qualify for an unreserved seat if their rank is good enough; unlike in the supernumerary scheme where girls can only occupy the 14–20 per cent of seats earmarked for them. 'One kind of reservation is N% reservation plus whoever can get in unreserved, and the other kind of reservation is minimum N% reservation. How is it not a reservation at all?' he wrote.

We became curious to understand Dheeraj's thoughts about reservations, so we met up with him one afternoon outside one of IIT Kanpur's many campus teashops. 'I think that with reservation, essentially you are making a statement that these people are not good enough, and that affects the self-confidence of people,' he said. Is there a better way? Yes, he says, suggesting a model where 'deprivation points' are awarded to students from marginalized communities to increase their chances of admission. This was something followed by JNU in Delhi (apart from reservations) until the new administration controversially scrapped the policy in 2016–17.

Such a point-based system sends a vastly different message, according to Dheeraj. 'What this says is that this set of people [marginalized] is actually performing better than the other set, but my exam is wrong and therefore I must find a way to admit them. With reservations, what we are saying is that women are not competitive, but inclusiveness and gender balance will somehow improve interpersonal culture, so we will admit them though they don't deserve to be here. So we are admitting them because of some other side effects. It's two completely different arguments. I am saying that they should be admitted because they are better and they are saying they should be admitted despite them being bad.'

Deprivation points, quotas, or supernumerary seats, whatever the preferred mechanism—the reason for the poor gender ratio in IITs comes down to the JEE exam itself. Never since its inception in 1961 has a girl topped the JEE Advanced (first known as CEE, then JEE). Not only that, it is extremely rare for girls to even make a presence in the first 100 ranks. 17-year-old Kavya Chopra, the top-ranked girl in the 2021 JEE Advanced, was placed 98th. Girls' underperformance in the JEE strikes as odd because girls typically do as well or better than boys in their final school board examination. Why this discrepancy? What is it about the JEE exam that suits boys better? The most commonly cited reasons for this involve lazy stereotypes about girls being unable to perform under pressure or being more built to 'mug up' facts than to 'apply concepts'. The more accurate explanation for the poor performance by girls in the JEE is that they are less likely to receive the years of financially, physically and mentally expensive 'coaching' that helped most successful IITians pass the exam.

The JEE is hailed as the epitome of meritocratic integrity, identifying the best young minds in the country year after year. This idea is problematic on many levels—the implied suggestion that our best minds are predominantly male is just one of them. Luckily, there is at least one study that seriously challenges this claim. In 2016, Ravinder Kaur who was heading IIT Delhi's Department of Humanities and Social Sciences was asked by the Director to lead a study on student performance at IIT. Unbeknownst to many, underperformance within IITs is a big concern and a mind-boggling one because these students are supposed to be the best and the brightest in the country. While everyone is quick to blame this on the 'quota' students, the fact is that even the general quota students underperform. 'It's a secret inside IITs. What gets emphasized is the few who get huge packages, not others who don't get jobs or get poorly paid ones,' said an IITian who did not want to be identified.

To shed some light on the phenomenon of underperformance, Ravinder and her colleagues used a sociological lens to analyse performance data spanning 13 years (2003–2015) of the undergraduates in her institute. 'While looking at this, we also considered the gender aspect and we found that irrespective of their [JEE] ranks, girls were performing one grade point better than boys.' Though an executive summary of this study was made, the study didn't end up being published and released to the public.[6] Nevertheless, Ravinder's study made a considerable impact—it became the most crucial shield against naysayers of the supernumerary scheme. It served as clear evidence that girls, once inside IIT, were doing better than boys, so by including more JEE Advanced qualifying girls, merit is not being diluted. However, the same people who are quick to bring up Ravinder's study to defend the scheme seem unconcerned or unwilling to address some big inconsistencies: Why are girls who do better in school unable to qualify for the IITs? And why are girls who do enter IITs doing better than boys who outranked them? There is an elephant in the room and it's asking—how effective is JEE as an entrance exam, really?

An article in *Forbes India* from 2012, put JEE among the toughest three exams in the world. It said, only half-jokingly, that the idea behind JEE was 'to weed out mediocre talent and pick only those who could solve S. L. Loney's trigonometric identities while brushing their teeth.'[7] The article went on to say, 'A JEE question wasn't just a question. It was designed to knock the intellectual stuffing out of the candidate.' The evidence from school exam results and Ravinder's study now prompt us to ask: Do our 17- and 18-year-olds really need to have their intellectual stuffing knocked out of them even before they join college?

Despite all these reasons to doubt it, there is great resistance to criticizing anything about the JEE. All the IIT Directors that we spoke to declined comment on the defects in the exam. A social scientist and former IITian, who did not want to be named, said, 'They talk about it but frankly, I don't see them doing anything about it. It seems like a proprietary thing, this great JEE of theirs. JEE seems to test for a particular kind of skill set that coaching helps [foster], but that doesn't necessarily help in creativity and empathy.'

Dheeraj Sanghi also believes that changing the JEE is next to impossible. 'JEE is a holy cow,' he said. According to him, any change in JEE is a suggestion that the process has been faulty all this time. And that is extremely risky for the IITs.

'If you say that this exam is really not finding the merit that it wants, then somebody who passed this exam last year asks 'are you saying I'm not meritorious?' Even a person who passed 20 years ago says this. Alumni of IIT raise ruckus every time there is a small change suggested in JEE.'

Why does the alumni's approval matter so much to the IITs? This sentence from an *Economic Times* report from 2019 might help you wager a guess: 'Alumni donations to the top five Indian Institutes of Technology (IITs) may cross Rs 1,000 crore by the end of this financial year, as the prestigious institutes have stepped up efforts to tap into this source of funding and well-placed former students are increasingly loosening their purse strings to give back to their alma maters.'[8]

Yes, it's all about the money.

The case of IIT perfectly illustrates how engineers and scientists cannot be expected to come up with the best solutions to sociological problems. There have been some questions about why the JAB sub-committee had no sociologists in it. Even Ravinder Kaur, the IIT Delhi sociologist who authored the famous internal study comparing the performance of girls and boys in IIT Delhi, was not part of the original sub-committee. The high stakes of alumni donations and the booming coaching institute industry further complicates things, confirming that diversity, beyond a certain point, is simply too inconvenient and risks making the big players unhappy. Instead, the issue is deflected citing fears such as 'dilution of merit'.

Even if 'dilution of merit' was a legitimate issue, it is only brought up selectively. Take the peculiar case of private pre-university (PU) colleges in Bengaluru, which have apparently been struggling with a surplus of girl students! Girls were consistently outperforming boys in the school examinations, so the colleges were becoming more and more female-dominated. The management of these colleges could not bear this. They claimed that there are already very good women-only colleges but not enough men-only colleges. Co-educational colleges must be gender balanced, they insisted. For the sake of 'equality', they implemented a higher cut-off mark for girls. This was first reported by mainstream media in 2010 but the practice has continued for over a decade. 'If there is no higher cut off, the college will have only girls. The higher cut off is to bring gender balance,' justified the vice-chancellor of one university, according to a report in the *Times of India*.[9] We see that when the ratio is lopsided in this direction, gender balance becomes a much better cause worth fighting for. But is it really about equality, or is it self-interest? It's possible that the management sees boys as better investments. After all, in the current social set-up, boys are viewed as more likely to become super-achievers and end up with high-paying jobs and perhaps support the institution financially in the future. While there was public outrage against PU colleges when this made news in 2018, it was interesting to note that the narrative centred on the unfairness of this on meritorious girls. Nobody seemed worried that the 'unjust' inclusion of lower-ranked boys would dilute the merit of the colleges.

When anthropologist Ajantha Subramanian interviewed students of IITs for her book *Caste of Merit*, she noted something interesting. When the underperformance of upper-caste students was being discussed, it was usually a matter of them 'having fun', whereas for students from marginalized caste backgrounds/students availing reserved seats, it became a matter of their intellectual capacities.[10] Similarly, when scientific misconduct claims were slapped on an upper-caste professor at an elite institute in Bengaluru, the reactions from their peers were mainly directed on the misdemeanours of the students involved in the work.[11] But when a Dalit professor at IIT Kanpur was picked on by his upper-caste colleagues for supposed plagiarism, he was accused, in an email, of being an 'unsuitable, uninspiring, unfortunate candidate', who was admitted through 'wrong means' and was 'mentally unfit to take the job'. The professors involved in this attack were later declared guilty of bad conduct and violating the SC/ST (Prevention of Atrocities) Act.[12] These examples reveal that the Indian science community is much more protective of merit in some cases than others.

Even if affirmative-action policies such as reservation are adopted, an institute's responsibilities do not end there. The terrible case of the casteist IIT Kanpur professors illustrates how beneficiaries of these policies can become targets for harassment. Unless there is unconditional support from sensitized colleagues, marginalized people whom such measures are meant to benefit will only suffer. Sadly, this is rarely the case. This explains why most women scientists we met were wary of reservation. They want to be seen as 'meritocratic' and are unwilling to bear the baggage that comes with such affirmative-action measures.

Diversity done wrong can get uncomfortable for the intended beneficiaries and we saw a range of examples of this. Award-winning computer scientist Sunita Sarawagi scored the second-highest CGPA (cumulative grade point average) in her whole batch during her undergraduate days at IIT Kharagpur. A mean feat in itself, the institute felt the need to reward her for being the best-performing woman. Sunita found this patronizing since there were only 12 women in her batch. 'For women, it can be demeaning to get this feeling that they have been chosen for this award because somebody wanted diversity.' A PhD scholar we met at a conference told us about a time when she was asked to give a talk at an outreach programme 'because I'm a girl and since they want gender equality'. She pleaded, 'Equality should be given to us because we are competent enough. Not because I bear this burden of gender equality. I'm happy that society wants India to become gender equal but we are talented too.'

We ran a poll on Twitter for anyone from an underrepresented group in science asking 'If you were hired/awarded a prize by a committee/jury who has said that they keep diversity in mind while decision-making, would you feel insulted?' 63 per cent answered 'yes'. This led us to Erin Leigh Howard, an undergraduate student of science at Western Washington University. Erin, who has autism, told us that special treatment doesn't necessarily insult them. They said, 'I wouldn't feel insulted if I was given an award. Just because my Google scholarship was for people with disabilities doesn't make it any less of a Google scholarship. However, I'd be wary if my disability was a reason I was hired. It depends how many others already work there.'

Various incidents have made Erin wary of disability diversity hiring initiatives and training programs; for example, in December 2019, the U.S. Geological Survey came under criticism for enrolling autistic individuals in research programmes without paying them as they did other participants.[13] Because of these prevalent attitudes, Erin is forced to ask themself 'Will I get paid? Will I get paid less than others? Are they assuming my abilities? Am I an exhibit?' every time an opportunity like this comes their way? They elaborated, 'If I was hired for my (lack of) gender or for my asexuality, I'd be OK with the initiative. However, people often assume my abilities rather than ask about them when they find out I'm autistic. When they do ask questions, it's often very personal and not related to what I'll be doing on the job. I also worry about becoming used as a free disability education dispenser, and not getting paid as much as my peers. If there were other disabled people who could vouch for them, OR if I knew I wasn't the only disabled person who'd be hired, I'd be more comfortable.'

However uncomfortable it makes scientists in our country, reservations work. V. Rama, a faculty member at the National Institute of Technology Warangal in Telangana, is one of the very few scientists we met who spoke openly about her own availing of reservations. Perhaps this is because Rama had a very unconventional trajectory. Coming from a family of modest means and stricken by polio, Rama told us in 2017 how she lacked direction and mentorship in her early years despite being academically brilliant. She was able to join the college as a lecturer with a BTech degree thanks to the reservation policy for Other Backward Castes. This does not mean that she had it easy. In fact, Rama had successfully overcome more challenges than the average candidate with higher qualifications. She had spent nearly four years pursuing a diploma course at a polytechnic college near her home, before finally enrolling for a BTech, which she could afford thanks to a scholarship. By this time Rama was already married. She had her first child in the first year of her BTech and her husband offered her no support. The couple eventually separated and Rama had to navigate single motherhood. Rama said, 'I was doing my BTech in the evening batch, so I worked in a government college during the day and studied from 5 p.m. to 9.30 p m. I managed to buy a bicycle and that helped my commute with my polio-affected leg. It was very difficult to go for higher studies.'

After her BTech, she decided to prepare herself for the UPSC exams to enter the civil service. For two years, she gave her all. Rama could only afford the UPSC training because there was a free batch for students from backward castes. While there was free accommodation and food for the men, as the only woman in her 100-student batch, she had to spend a whole month negotiating with the management before any accommodation was provided to her. Unfortunately, Rama could not crack the famously competitive exams and she could not afford to sacrifice another year for this. In spite of all this, she credits these months for building her character and strength.

'I can't tell you how much I struggled, but it was a wonderful period in my life. I became more confident. I think every woman should have some coaching in civil services. Unfortunately, engineering students know nothing about politics or our Constitution, they have no social awareness.'

Rama may not have had access to a job at her institute if not for the reservation policy but it's just as important to emphasize that without reservation, her colleagues and students would have missed out on the chance of working with someone with such a rich life experience.

Inspired by the popularization of gender equity in STEM topics in the developed world, Indian scientists have, in the past decade or so, become emboldened to agree with their Western counterparts that 'affirmative action' is a necessity. Affirmative action refers to policies that offer preferential treatment for minorities and women as an attempt to compensate them for being denied opportunities of advancement due to past and present discrimination.[14] Yes, that sounds a lot like reservation, doesn't it? Yet, the same scientists continue to distance themselves from reservation, an affirmative-action policy that India adopted much before countries like the USA.

The highly patriarchal Indian science community has a long way to go in ridding itself of its ignorance and prejudices about reservation. Notwithstanding this unfortunate scenario, it's important to realize that the gender-equity movement in Indian science isn't as new as many of us think. It is built on the strong foundations of decades-old struggles, surveys and in-depth studies led by women researchers themselves, sometimes with the support of their institutions and the government. This is what our final chapters are about.

Part IV

Our Science Culture Must Change

17

Nudges from the Top

In 1994, Seema Pooranchand left the prestigious PRL in Ahmedabad with a PhD in hand. It was the culmination of an arduous seven-year-long research journey for the young astronomer who had taken up the challenge of developing an instrument to study galaxies. At PRL, Seema also met her life partner, Syed Maqbool Ahmed, who was her senior at the lab. The couple departed for the USA, where Syed began a project with NASA's Jet Propulsion Laboratory and Seema took up a research position at Caltech. This was the high point of Seema's scientific career. The couple's son and daughter were born after their return to India. Both suffered from health issues in their childhood, and their daughter was diagnosed with muscular dystrophy at the age of seven. The lack of a support system subsequently pushed Seema out of mainstream science. She clung to astronomy by coaching students, delivering public lectures and motivating young people towards the field. But by the time things were stable enough to consider resuming research, Seema realized that employers were no longer interested in her. Such losses are depressingly common in Indian academia, but Seema's story is notable for the fact that she returned to research after an 18-year hiatus.

Seema's comeback was facilitated by one of the most popular government schemes for women in science, called WOS-A. When this scheme was launched in 2002 by the DST, it was simply called 'Women Scientists Scheme'. It was intended to provide opportunities to women scientists by giving them grants worth Rs 10–15 lakh to use over three years. Interestingly, the aspect of the scheme that really took off was the special 'Break in Career' category targeted at scientists like Seema who had taken a few years off, often due to motherhood or family responsibilities. The scheme was an instant hit—the DST reportedly received over 2000 applications in the first year alone. Later, the 'Break in Career' category would become a separate scheme called WOS-A. Until it was remodelled yet again in 2023, the beneficiaries of WOS-A were provided with a research grant of Rs 30 lakh over three years (for PhD holders), which included a fellowship as well as travel, equipment and other costs. Applicants were required to be between the ages of 27 and 57, and the 'break' was required to be at least two years long.

Seema knew it would take a damn good proposal to compensate for her 18-year break from active research, so she meticulously prepared one. To her surprise, she found that her interviewers were actually excited to hear about her unusually long break. 'They said it was a great thing [to want to come back to research]. It was a very positive thing for me,' she shared.

Though the DBT already had special awards for women since 1999 and had launched initiatives such as a biotechnology park for women in Tamil Nadu in 2001, the DST's first big acknowledgement of the gender gap was the WOS-A scheme. Since its launch, WOS-A has given thousands of women scientists a second chance. It is repeatedly brought up as a positive example of governmental support for women in science. In a Twitter thread posted on International Women's Day in 2019, the then Union Minister of Science and Technology, Harsh Vardhan, boasted that 'Over 2100 women scientists have been brought back into mainstream science under Women Scientists Scheme', while he did not clarify where these numbers were from, he was presumably referring to the number of beneficiaries of the WOS-A scheme until then.

Beneath the facade of the success stories, however, is the grimmer reality of WOS-A and similar schemes introduced to benefit women in science. 'WOS-A has been a lifesaver for many, but it is meant to be a bridging programme and this is not happening,' summed up a senior scientist who did not want to be named. Indeed, the minds behind the WOS-A scheme had envisioned that, over the course of the three-year fellowship, worthy beneficiaries would be absorbed as permanent faculty at their host institutions. However, this rarely happens. So at the end of the three years, most WOS-A scientists are back to square one—jobless, resourceless and, as we gathered from our interviews, more demotivated than ever. The DBT's BioCARE programme, also open to women with career breaks, lasts for five years and the employment situation here is similar. According to a 2019 press release,[1] 165 unemployed scientists had received the grant since 2011, of which only 28 had procured permanent employment.

In 2020, we caught up with an official from the DST's women-in-science division. The official admitted that the DST was well aware of their scheme beneficiaries' challenges in securing long-term employment. However, they said that there was a 'lack of data'. According to the official, who was reluctant to be named, a third-party review of the scheme had been instituted and a report was expected in six months. However, two years later, we were still unable to trace the results of this review. When things are this opaque, it becomes impossible to say how effective the WOS-A scheme actually is for women in science in the long run.

The other serious issue plaguing WOS-A is the delayed disbursal of funding, which is an alarmingly common problem faced by researchers in India. For WOS-A scientists who only have three years, such hold-ups can sound the death knell for their research goals. The funding agencies blame the delays on shoddy paperwork submitted by the scientists or the host institutions. However, the quantity and quality of the complaints suggest that the situation is more serious than that. When we first interviewed Seema in 2017, she herself was in the middle of such a bind. She was five months into her second year of the fellowship but had not yet received the second instalment of the grant. 'When I called them up, they said they had not updated my file. That's how it goes. Still, I'm trying to do my work,' she said, with a pained smile.

'It's okay for me since my husband is working but . . . we worked so hard to get it [the fellowship]. It took me one year to write the proposals, it's disappointing. People say this happens with the DST. This is quite late though. I want to request them to look into this, so that research can go on smoothly.'

Seema continued looking for a full-time position but nothing had worked out for her when we last spoke in 2022. She was keeping herself busy by organizing workshops and astronomy outreach events in her neighbourhood and virtually. Meanwhile, her daughter Khushbu completed her master's in health psychology and both mother and daughter had a keen eye out for good jobs. Seema's WOS-A project concluded in 2018, but it was only in late 2021 that she received the final instalment of her grant money.

A long way off, at Sister Nivedita University in Kolkata, Rupali Gangopadhyay shared with us her own bitter experience with WOS-A.[2] Being a WOS-A beneficiary had left the chemist depleted and thoroughly disillusioned. The whole three-year period, according to Rupali, was fraught with irregularities. Ultimately, she was not able to purchase the equipment and software that her research demanded and her work was left incomplete. She furiously recalled numerous unanswered phone calls and emails to the DST. On one occasion, during a 'project monitoring' meeting by the DST officials, Rupali and other WOS-A scientists stated plainly that the reason their research was not progressing was that they had not received their grant. Instead of receiving an explanation, they were harshly told that 'this was not the place to discuss [this matter]'. Another time, Rupali bumped into a representative of the DST during an event at her institute in Kolkata. Recognizing her, the official promised her that the money would reach her soon.

'Nothing happened after that and there was no communication. I sent an SMS to his number, which he had shared with me. When I gave him a call, he told me "just because I had given my number, doesn't mean you can call me".'

The repeated humiliation prompted Rupali to write about her experience in a personal essay that she published online. In it, she bemoaned how the poor execution of her WOS-A had reduced someone like her, a scientist with dozens of articles in international journals, a highly prestigious postdoctoral fellowship and a winner of several national awards, to the position of a '*beggar*'. 'All I am asking for,' she stressed, 'is to be reimbursed for the money I have already spent on my project or the salary for the months I worked unpaid.' The chemist eventually made peace with her teaching job at the university and has all but given up on her quest for a research job.

Rupali is sure her experience is not an outlier. Ever since her essay and a *Current Science* report, which she co-authored in 2016 outlining the defects and possible solutions of the WOS-A scheme,[3] she has been contacted by several others who have faced similar situations regarding the release of grants and reimbursements. Was there any response from the DST? 'Nothing,' she replied. When we interviewed her nearly seven years after the completion of her project, she was still owed an amount of over Rs 6 lakh.

Looking back, Rupali has two pertinent questions: First and foremost, how can such schemes hope to mainstream highly educated women scientists like her if they continue to mistreat, underpay and dismiss their scientific ideas? Second, is it really impossible for funding agencies to treat the beneficiaries of their own awards with respect and maintain an open channel of communication with them?

The answer, she writes, is blowing in the wind.

In 2014, the DST combined its programmes targeted at women scientists under a division named KIRAN. An acronym for Knowledge Involvement in Research Advancement through Nurturing, the KIRAN division includes the WOS-A scheme,[4] as well as the WOS-B and the WOS-C schemes, which provide grants for projects related to 'Science & Technology interventions for societal benefit' and training for intellectual property rights management, respectively. Also under this umbrella is CURIE (Consolidation of University Research for Innovation and Excellence in Women Universities) aimed at developing the infrastructure of women-only universities.

Among the newer schemes launched by KIRAN is the 'mobility scheme'. The mobility scheme seems to be a response to the familiar situation of women scientists being forced to leave their existing permanent jobs for reasons such as their spouse's posting, the health of a family member or the education of their children. The mobility scheme, according to the DST, is designed to 'act as filler while searching other career option at new place [sic]'. There is very little information on the mobility scheme available in the public domain. Even our source at the DST was unable to satisfactorily explain why the scheme, which was announced in 2016, had yet to take effect in 2020. Yet again, it is not apparent how this so-called 'landmark' scheme achieves the intended impact.

It becomes even more challenging to be optimistic about the DST's initiatives when we examine the language that they use to describe their programmes. The department comfortably and unquestionably perpetuates, normalizes and even glorifies patriarchal family dynamics, where it is assumed that the husband's job is more important. This is how the mobility scheme was described in the DST's advertisement:

> The initiative intends to provide a harmonious environment during early phases of women scientists where they would like to stay active in research in addition to attending and fulfilling other responsibilities in the domestic front.[5]

The tone and nature of some of the DST's schemes to perpetuate stereotypes have been questioned by several scientists. Acclaimed physicist Rohini Godbole said in an editorial letter for the journal *Current Science*:[6]

> . . . *Such schemes should be meant for couples, with either of the partners doing the relocation or the redefinition. Another example is a creche. It is considered to be necessity if the number of women employees is above a certain minimum, the inherent assumption being that it is only the women*

employees who need a creche, whereas in fact a creche may
help a male employee manage dual careers with his partner. It
is important to think of these and other steps we take in future
in a 'gender neutral' way.

As 'women in science' becomes a stronger buzzword and
the tolerance for the gender gap reduces among the STEM
community, all eyes are on the DST to see whether they will
rise to the challenge. Hopes had risen with the announcement
of their new scheme GATI, short for Gender Advancement
for Transforming Institutions, which was announced on
National Science Day 2020 by the then President of India
Ramnath Kovind. It is a charter consisting of ten principles that
institutions will be encouraged to adopt and, thereby, declare
their commitment to 'support diversity, inclusion and the full
spectrum of demographic talent for their own success and
progression.'[7]

GATI was inspired by the UK's Athena SWAN (Scientific
Women's Academic Network) charter, which was established
to encourage and recognize the commitment to advancing the
careers of women in STEMM (science, technology, engineering,
math and medicine).[8] Since its inception in 2005, the Athena
SWAN framework has been adopted and adapted in Australia
(SAGE Athena SWAN), the USA (SEA Change), Canada
(DIMENSIONS) and now India (GATI).

In the summer of 2021, the GATI pilot was underway. In
the cockpit was Pratibha Jolly, physicist and former Principal
of the University of Delhi's women's college Miranda House.
When we caught up with her the same year, she was bang
in the midst of mission mode—harried but hopeful. 'This
is a DST project being rolled out by KIRAN and the British
Council is on board as well,' she said, 'and as the PI for the

project, I am developing the framework for assessment and accreditation.' Though the charter had already been drafted, Pratibha was still in the process of building her team, which would consist of a social scientist, a policy researcher and a data scientist. Together, they would create a comprehensive framework that would guide the pilot institutions. An open call for applications from institutions to be part of the pilot had yielded 148 expressions of interest, of which Pratibha's team had the tough job of selecting about 30.

The 30 institutes finally chosen to be in the pilot included IITs, research institutions and central, state and private universities. They were to be assigned UK institutions chosen by the British Council as 'partners' to handhold and mentor the Indian institutes through this new programme. Once onboarded, the pilot institutes would have about a year and a half to gather their data, according to Pratibha. 'That's the easy part. They will then have to decide how and what they should be doing . . . an action plan,' she said.

A crucial player in the GATI scheme is the 'nodal person' appointed for each institute. The nodal person would be the point-of-contact and the coordinator of all GATI-related activities, including data collection as instructed by the framework. Eventually, the participant institutes would be assessed and accredited based on their performance. While we write this during the summer of 2022, it is still too early to say what GATI's grading system will look like, but Pratibha has hinted to us that it may not be the same gold, silver and bronze awards format as Athena SWAN.

'We cannot expect any significant change in numbers in these institutes in the next one-and-a-half to two years, especially not now with the pandemic. The focus [for the pilot] will be on where we stand, what we can do, what we should do and how we should do it. The work has to be on a continuum. It has to be a lifetime project,' she stressed.

For Pratibha, it is not accurate to call GATI an Indian version of Athena SWAN. She explained:

'It is not possible to map Athena SWAN onto India. We are two different countries and our frameworks are entirely different. The only thing is, there is much to learn from . . . the way they manage the whole thing.'

When we compared the GATI and Athena SWAN charters, we did notice some stark differences but we were not convinced that these were a reflection of the uniqueness of the Indian context. Instead, GATI seems rather tepid, compared to its British counterpart. Two key points of concern stood out to us.

One, where is the 'commitment'?

While almost all of Athena SWAN's 10 charter principles begin with the pledge 'We commit to advancing/addressing/ tackling/removing . . .' most of GATI's principles opt for non-committal verbs such as 'We acknowledge/appreciate/ realize/recognize/are deeply concerned . . .' The UK's charter also demands a much higher level of accountability from their institutes and specific problem areas are brought up. For example, they pledge to address 'the negative consequences of using short-term contracts', tackle the 'discriminatory treatment often experienced by trans people' and the 'gender pay gap'. In contrast, some of GATI's principles read as vague and obtuse. For example, Principle #10 is 'We believe that mainstreaming, assimilating and sustaining positive impact policies and actions for gender advancement will bring transformative changes in the overarching climate and socio-cultural ethos leading the institution towards distinctive excellence.'

But Pratibha believes that the GATI charter principles are a reflection of the Indian scenario and that the onus is on the institutions to use the charter as best as they can to create change in their own respective atmospheres. GATI, she repeatedly insisted, is just a 'nudge'.

'Our goal is not to go and solve problems for institutions. It's not going to be a top–down role. We won't be telling them to do this or do that. It will be a community-based project. They have to do it themselves.'

The second point of concern is more pressing: Does GATI know what it means by 'diversity'?

Principle #1 acknowledges the importance that institutions provide equal opportunity for the 'broad diversity' of stakeholders. Sayantan Datta, a queer and trans science writer and commentator on policies in Indian science, shared their insights on the charter. 'When they say "broad diversity" do they include queer, trans, Dalit, Bahujan, Adivasi and disabled persons in science? Or if GATI focuses more on the axis of gender, then what/who is/are the broad diversity?' Sayantan is also concerned about how anybody could possibly monitor this. '"Equal opportunity" is not a simple term and without affirmative action, no opportunity can be truly equal for the marginalized.' Principle #3 seems to answer part of Sayantan's question, with the phrase 'people of all genders', offering a glimmer of hope that the DST was aware of the non-binary nature of gender, but this hope is instantly crushed. Subsequent principles revert to focusing on women alone.

It was even more unnerving to see that in the application form for institutes to GATI's pilot programme, only three statistics were requested: male, female and total. Data on other sexes and genders who may be at the institute is not explicitly encouraged. This appeared to be against the Supreme Court's NALSA judgement, which mandates all documents provide the option of 'third gender'.

When we asked Pratibha if GATI is equipped to accommodate trans and non-binary people in science, she said that this was an issue she was well aware of. She admitted that while issues of transgender persons and intersectionality were not explicitly mentioned in the GATI charter (as they are in the Athena SWAN charter), Principle #2 does mention wanting to nurture the talents of 'all' people. 'The Indian system is very complex at this moment,' she said. She concluded by stating that GATI may very well evolve, especially with the drafting of the new Science Technology and Innovation Policy (STIP 2020) document. 'If STIP is notified quickly, we will adopt it too.'

For those who are loath to put all their apples in the GATI basket, there's another equity policy that is seeking attention. This one takes the form of a chapter within the draft document of the STIP 2020, which was developed jointly by the Office of the Principal Scientific Adviser to the Government of India and the DST.[9] As of June 2022, this draft policy document was still awaiting the ministerial nod that would make it official. When the STIP 2020 draft was released to the public, it was immediately heralded as a revolutionary step in policymaking. There were many reasons for this. Notably, the drafting process was more transparent and inclusive than most other policy undertakings. It included periodic live online meetings with various stakeholders and a website that was fairly active and easy to navigate. The team was also active on social media.

What resulted was a comprehensive vision of Science, Technology and Innovation, including chapters dedicated to science communication and equity, subjects which are otherwise pushed to the periphery of science in India. The language in the Equity & Inclusion (E&I) chapter make it evident that it has been drafted by people who actually understand many of the nuances of the gender gap in India. For example, it breaks away from the usual self-aggrandizing rhetoric with the admission that all is not well with Indian science. It refers to 'an absence of an inclusive culture', 'inadequate incentives', the need for gender-neutral policies and a deeper understanding of social barriers. It also acknowledges the LGBTQ+ communities.

However, once we get over the dazzle of 'woke' sentiments so rarely expressed in government documents, it becomes evident that STIP 2020 has serious problems. For example, it places a lot of emphasis on full-time-equivalent researchers, a category which excludes most women, transgender, Dalit, Bahujan and Adivasi people in science in the country as large numbers hold temporary contracts. A more shocking aspect of it is its use of the term '*divyangjan*' to refer to people with disabilities despite it being rejected by many disability groups, including the United Nations' Committee on the Rights of Persons with Disabilities.[10] The word literally translates to 'people with divine bodies', yet its usage is consistently advocated by Prime Minister Narendra Modi who once said '*Divyang* people have special powers'. Usage of this term is considered extremely patronizing and derogatory and takes away from the reality of the lives of disabled people but it seems the policymakers behind STIP 2020 were ignorant or indifferent about this. Policy researchers Shambhavi Naik and Priyal Lyncia D'Almeida, based at the Takshashila Institution, found STIP 2020 so inadequate that they decided to rewrite some of its chapters, including the E&I one. 'When I first read it, I thought "this" is not policy. Yes, policy documents should have a vision and STIP 2020 has that, but we need some contours on how this vision will be implemented!' said Shambhavi when we caught up with the duo to discuss their motivations. What stood out to both was that this was a science policy document with no data. 'We rewrote the chapters to demonstrate how a policy document can and should be structured.'

Despite their flaws, GATI and STIP 2020 signal a shift in the sensibility and direction of women in science policies in India. Unlike previous schemes and policies that aimed to 'help' women resurrect their scientific dreams, these initiatives emphasize the need to practise gender equality for the benefit of science itself.

Compare this excerpt from the preamble of WOS-A:

> Through this endeavour of the Department, a concerted effort would be made to give women a strong foothold into the scientific profession, help them re-enter into the mainstream and provide a launch pad for further forays into the field of science and technology.

With principle #10 in GATI:

> We appreciate that mainstreaming, assimilating and sustaining positive impact policies and actions for gender advancement will bring transformative changes in the overarching climate/socio-cultural ethos/ecosystem leading the institution towards distinctive excellence.

Even if GATI and STIP 2020 are eventually implemented, will they help science in India achieve gender equality? It's important to remember that neither is strictly enforcing anything. So we are relying on the scientific community to willingly and quickly turn over a new leaf and change their old habits, something that they have not been able to achieve for decades. Such a level of reform calls for exceptional leadership from the powerful people at the top of Indian STEM. Leadership that reflects integrity and empathy. Leadership with the humility and courage to recognize wrongs, practise and demand accountability, even from one's own ilk. Leadership with the will to make corrections even when it means having to let go of some privileges. Do our leaders have what it takes?

This true story may help you answer this for yourself.

Towards the end of 2020, the Indian science community on Twitter was flabbergasted by the announcement of a bizarre 'debate' that was scheduled to take place during the upcoming government-organized India International Science Fest. Titled 'Do women make better leaders?', the debate would pit two teams of scientists against each other. The first team would comprise early-career women scientists who would have to speak *against* the topic (i.e., Women do not make better leaders), and the second team of early-career men would speak for the topic (i.e., Women do make better leaders). BiasWatchIndia, an initiative by scientists Shruti Muralidhar and Vaishnavi Ananthanarayanan, promptly posted a Twitter thread discussing the various reasons it was inappropriate for the DST and the CSIR to endorse a discussion about the aptitude of women leaders.[11] Moreover, 'Having young women argue against this is also morally wrong—the optics for young women is horrible!' they pointed out. Despite many days of outrage by scientists, students and communicators of all genders and across seniority levels, the event went on as intended. The few government officials who bothered to reply were defensive in nature. Instead of demanding action from the organizers, these officials urged the offended Twitteratti to be more polite and empathetic towards the supposedly well-meaning organizers. Organizing such debates could help us 'improve gender parity', tweeted Shekhar Mande, the then Director General of CSIR.

However, the organizers were rattled enough to follow up with a change in the title of the debate, though they offered no explanation or apology. The new title would be 'How to promote women leadership in scientific institutions? Systemic changes are necessary'. On D-Day, many of the protesters, including us, were present at the live virtual event in the hope that there would be some acknowledgement of the discussions that preceded it. This was not an unreasonable expectation, after having offended so many women in science. However, we were in for a rude surprise. The debate went on as originally planned. The title change, it seemed, was merely an attempt to shut the nay-sayers up. The only references to the outrage were snide comments by the moderator of the debate, a motivational speaker. The moderator brushed off the controversy as a case of a bad sense of humour and spouted fatuous and erroneous pieces of 'wisdom' such as 'No one can make you feel inferior without your consent' and 'Merit will be manure, merit will be the fertilizer'. The worst was yet to come. During the 'debate', the moderator went on to refer to queer-trans people as 'LGBTQ-ABCD whatever'.

In their searing critique of the debate,[12] Sayantan Datta wrote this:

> . . . to me, as a queer-trans person in science, her words provide a glimpse of the cruel reality of everyday living in science. It's true that as a queer-trans person, being in science is a lonely experience. Most people around are not queer or trans, and are often ignorant about issues of queer-trans people. However, this sort of proud demonstration of the disregard and ignorance that science as a discipline harbours for queer-trans people is, I argue, harmful to the cause of inclusivity in the sciences and to the entire act of science itself.

. . . the 'ABCD, whatever it is' also puts something more harmful on display: how privileged, cis-heterosexual people don't care about queer-trans sensibilities. As a queer-trans person attending the debate, I was triggered by such a reference and my mental health has been affected by it.

. . . Broadcasting such messages will discourage young queer-trans people from entering science, thus taking away many new possibilities and perspectives they could have brought to the discipline.

The silence from some of the biggest leaders in science who were among the live audience that day was truly gutting. Nobody seemed perturbed by the grossly sexist way the event was turning out. None of our leaders, including popular figures such as then CSIR head Shekhar Mande and then Principal Scientific Adviser K. Vijay Raghavan, displayed any solidarity that day. And a few weeks later, it was as if this episode had never happened.

Can the same leaders who could not even stand up to an outright misogynistic and transphobic debate suddenly develop the will to make policies like GATI and STIP 2020 work? We have our doubts.

18

Tremors from Within

The various government schemes for women in science, the GATI charter and the draft STIP did not emerge from a vacuum. Whatever their flaws, they are the fruits of a two-decade-old (at least) movement for equity in Indian science led by vocal and determined women and trans persons in science, who have been consistently shaking things up from within their institutes, academies and beyond. Rohini Godbole is one of them.

In an autobiographical essay published in the book *Lilavati's Daughters*,[1] Rohini recollects an incident from 1989. During one of her travels, she bumped into a Japanese scientist who was blown away upon realizing that she was the author of a well-known study in the field of elementary particle physics. 'He bowed down to me in the middle of Frankfurt Airport and said "I respect that work!". I must say it did wonders to the self-confidence,' she wrote. At the time, Rohini was living a 'double life' in her own words. She was shuttling between the University of Mumbai, where she was a lecturer, and the TIFR, where she was doggedly pursuing research. Her perseverance paid off when, 12 years into her teaching job, she was invited to the historic IISc as a professor. Over the next 25 years, Rohini would go on to establish herself as one of the most iconic particle physicists from India. Along with German physicist Manuel Drees, she successfully predicted an important phenomenon that would aid the design of the next generation of linear electron–positron colliders. Her work would eventually lead experimental physicists to new avenues of particle physics research, an impact that gave Rohini unparalleled satisfaction and joy.

In 2021, Rohini was conferred the *Ordre National du Mérite*, one of the highest civilian honours bestowed by the French government. This was in recognition of not just her scientific achievements but also her commitment to the cause of women in science. Indeed, Rohini is one of the few prominent names in contemporary Indian science to have addressed the gender gap in a meaningful and sustained manner.

Growing up in a family of dynamic women and being educated in schools and colleges where she was routinely a top performer, Rohini had, for much of her life, never considered the idea of women doing science as anything out of the ordinary. 'We were used to seeing women doing housework as well as following their dreams—dreams of learning, not necessarily jobs,' she said, when we caught up for an online chat in 2021. Rohini became aware that gender bias was an issue in 2002 when she was asked to deliver a talk at the First International Conference on Women in Physics, held by the International Union of Pure and Applied Physics (IUPAP) in Paris. She had been invited as the Asian 'success story'. She said, 'It was the first time I thought about the subject. Not that I hadn't experienced difficulties earlier, but I hadn't thought about whether it could be because of my gender. While preparing for the talk, I realized that there were things in my own past that may have been influenced by my gender. I also realized that for every one of us who succeeded, there could have been many that did not.'

The conference in Paris fired Rohini up, and she began looking for opportunities to ignite a similar consciousness in her own country. At a meeting of the IASc, Rohini managed to get a two-hour slot for a discussion on women in science. Few people showed up, and those who did were not entirely convinced.

'People were looking around and saying 'Do we really need to discuss this? Do we really have such a small number in the academy?' I remember sitting there with the academy book in my hand, turning the pages and counting. When among 900 fellows, you see so few [women], that's when it hits home . . .'

Being an elected fellow of the IASc and INSA academies, Rohini had the reach to make something happen. She, along with a few others, convinced the then INSA President M. S. Valiathan to identify 'Science Career for Women' as a 'thrust area for investigation'. This led to the constitution of a committee of nine scientists, chaired by nutritional biochemist Mahtab Bamji, who was an early and important contributor to improving the representation of women in Indian science. Rohini was part of this committee. Over the next year, the committee, in partnership with social scientists at Shreemati Nathibai Damodar Thackersey (SNDT) University in Mumbai, produced a comprehensive report titled 'Science Career for Indian Women: An examination of Indian women's access to and retention in scientific careers'.[2] This was the first official study on women in science sponsored by the Indian government's DST and it was released in 2004. In Rohini's words, 'This was the beginning.'

It's not as if the DST was ignorant of the gender gap until then, but they had so far gotten away with women-only schemes and awards, which seemed more altruistic in nature than based on any evidence of bias. The INSA report was a much-needed wake-up call. In it, the committee used data from the UGC to point out a unique aspect of the gender gap in Indian science: While the problem in the West was that schoolgirls did not choose science when they went to university, this was less of an issue in India. Many children did, of course, drop out of science during school, but the gender gap at this stage was not stark. The more disproportionate attrition of women was happening post-higher education when it was time to find jobs. The situation stayed bleak for women who managed to get jobs but were looking to advance their careers. The INSA report also included a survey undertaken by SNDT University's social scientists whose results showed that gender insensitivity and discrimination in the workplace were real problems. The survey also inferred that the study of science by women was a privilege of those from urban areas and upper castes. Evidence of caste-based harassment emerged from the interviews they conducted. For example, the report notes:

> The upper caste lab attendants referred to a Dalit woman professor as Bai, whereas the senior male professors were called 'Sir'. While *Bai* when literally translated refers to a lady, in common usage it means a woman domestic help. It was obviously a way of getting back at the caste of the woman professor by the upper caste employees.

The document ended with a set of recommendations to improve the status of women in science.

The INSA report created a stir strong enough for the DST to respond with the formation of a task force. This task force's objective was to enable the implementation of the recommendations of the INSA report. Over the next three years, the task force conducted 10 meetings across the country with teachers and students of science and also solicited feedback from the public via advertisements.

Meanwhile, in 2003, the IASc academy formed its own committee on 'Women in Science' (WiS). The IASc committee, chaired by Rohini, came up with a set of 14 action points, followed by a panel to carry out those recommendations. The IASc's WiS panel focused not on studies as the INSA had done, but on on-ground activities such as seminars. 'These seminars would feature women scientists talking not just about women in science, but about top-class research,' said Rohini.

The IASc's most popular contribution was *Lilavati's Daughters,* a collection of 98 biographical and autobiographical essays about women scientists in India in the past as well as in the present. *Lilavati's Daughters* is largely a labour of love for Rohini and her friend and colleague Ram Ramaswamy. The title is an ode to mathematician Bhaskara's 1150 AD treatise. *Lilavati,* in which he writes about math problems in the form of poetry addressed to his daughter Lilavati.

Lilavati's Daughters came out in 2008 and created a lasting impact on the scientific community. which, until then, was starved of resources and stories about women scientists in the country. These essays shone the light on a slew of talented voices in science and also made talking points about common challenges faced by women in their scientific trajectories. It was subsequently translated into several Indian languages and also edited into a version targeted at younger audiences, titled *The Girl's Guide to a Life in Science.*

Another significant event took place the same year. The Mahtab Bamji-led DST task force organized a first-of-its-kind national conference of women in science on International Women's Day at Vigyan Bhavan in New Delhi. The conference featured an audience of over 1200 women scientists, teachers, students and entrepreneurs—with expertise ranging from cell biology and brain function to climate change and aeronautics—as well as high-profile politicians, including the then President Pratibha Patil. Besides an exhaustive range of technical sessions, there were also sessions on policy issues, opportunities and issues such as sexual harassment in academia.

This conference generated some criticism from scientists who questioned the effectiveness of its women-only nature. In a report in *Current Science*, scientists Vineeta Bal and Vinita Sharma wrote:[3]

> The only time a few men were present during the two-day conference was at the inauguration. In fact, during an interactive session, a pointed comment was made about the absence of men, with women wondering whether they were talking amongst themselves, and whether talking to the converts has any relevance at all . . . While such conferences serve the purpose of providing exposure to competent work done by women scientists and technologists to the world, in general, absence of male colleagues meant that women's work and achievements went unnoticed by male colleagues, competitors and bosses . . . Absence of male colleagues from this conference, thus defeated part of the purpose of the showcasing effort. The lesson to learn from this effort is not to organize women-only conferences, but to strive for a near-equal representation of both the sexes

as speakers and participants, even if that has to be achieved by affirmative action!

Criticism notwithstanding, something significant happened at this conference. Then Union Minister for Science and Technology, Kapil Sibal, came on stage with some good news. He promised financial support to establish crèches at all the DST-funded institutes; flexible working hours and work-from-home options for women scientists with young children; women-specific research grants and on-campus residential accommodation for women scientists. He emphasized that all scientific departments would need to ensure that these measures are implemented. These were all recommendations that had been made in the INSA report.

That should have been a major breakthrough in India's movement for women's equality in science, but it wasn't. Rohini remembers the minister's announcement during the conference. She recalled that it led to the formation of the Standing Committee of the Government of India for Women in Science in 2009. However, the committee, which she was a part of, never met. The announced promises weren't converted to a ministerial order and, thereby, did not materialize.

The DST task force was not idle at this time. Since its formation in 2005, it had been going to great lengths trying to reach all corners of the country. The group published a report in 2009 on the basis of feedback received in the ten meetings they held across the country. Titled 'Evaluating and Enhancing Women's Participation in Scientific and Technological Research: The Indian Initiatives', the report included chapters on work done in the past, an update on the status quo (five years had passed since INSA's report), as well as another list of recommendations.[4]

Evidence of systemic issues was piling up with every study and report being published, and by the end of the aughts, the atmosphere was thick with the anticipation of change. Fresh from the success of *Lilavati's Daughters*, Ram Ramaswamy and Rohini Godbole often spoke with each other about what more was needed. They knew that *Lilavati's Daughters* was just anecdotal and needed more analysis. Similarly veined discussions between natural and social scientists led to the commissioning of an ambitious survey that the IASc undertook along with the NIAS. According to Rohini, who co-authored the resultant report with Anitha Kurup, Maithreyi R. and Kantharaju B., the intention of the survey was simple: to find out why women leave science. For this, they would need to first build a sizable database of women with PhDs in science, including a substantial number of women who had left science. Rohini had a strategy in mind:

'I thought the best way to find women who left is via their advisors. So I sent an email to all fellows of the academy, requesting a list of their current and past women PhD students and what they are doing today. I must tell you I didn't get too much traction with this simple request. It took quite some perseverance.'

The team from IASc and NIAS managed to create a database with nearly 2000 women with PhDs in science, engineering or medicine from research institutes, universities and private and government-owned industries. The results of their study, published in 2010, were highly illuminating.[5] It provided evidence for unspoken suspicions, upturned myths and revealed surprising realities about the gender gap in science in India; all backed by something that all policymakers claim to love—data.

Arguably, the biggest achievement of the study was its refutation of the lazy assumption by policymakers that societal/

family responsibility was the singular factor behind the attrition of women from science. We will elaborate on this in the next chapter.

Embedded in the IASc–NIAS study were critiques of existing policies, especially ones that perpetuate stereotypical gender roles and stigmatize women. The authors pushed for policy-level changes. They recommended more transparency in selection and evaluation; a targeted time-based recruiting system at institutes with poor representation of women; mandatory disclosure of gender breakdown in all organizations; compulsory composition of one-third of women members on committees and hiring of spouses in the same organization. Rohini and her co-authors also recommended that WOS-A-like schemes be modified to offer more long-term working opportunities for women who are returning after a break. They reminded policymakers that the issue was complex and still relatively unstudied, so the periodic review of processes was crucial to make sure everything was on the right track.

After this flurry of efforts by natural and social scientists to provide a solid bed of resources and data, there was great expectation for some sort of action. Something, surely, had to happen. Sadly, the response from institutions and government leaders did not go beyond tokenistic gestures to appease women scientists. More reports trickled in over the next decade, but by 2020, it seemed the academies had lost their momentum and stopped pushing.

'In India, our policies are not too bad—the problem is that often people are more interested in starting new initiatives than in enforcing existing policies,' said Shobhana Narasimhan, a physicist studying computational nanoscience at the Jawaharlal Nehru Centre for Advanced Scientific Research, who has served on many of the WiS committees and also the DST's task force for women in science. Shobhana is frustrated every time she sees the government launch new schemes with great fanfare rather than revise and review the efforts that already exist.

'We don't have a pay gap (i.e., in academia, men and women who perform the same job are paid the same salary), we have rules mandating crèches and our maternity leave is reasonably generous. But no one checks if a place has a crèche, what the quality of it is, and who is using it. Maternity leave ends up being used as an excuse not to hire women. Spousal hiring, age limits, these things continue. No one does or says anything. There is a lot of complacency and self-congratulation,' she said.

Rohini chalks it down to apathy: 'I don't think people even read the [IASc–NIAS] survey. We need more support from the men in the community and more commitment. They have to realize that this is not just for women, it's for the benefit of science.' While Rohini is unsure whether or not the situation will change, developments in 2019 have given her dream of equality new life. She led the team that came up with the E&I chapter of the STIP 2020 draft, and it's something she is cautiously excited about.

'The drawback of previous initiatives is that they were only for women. The measures were voluntary and there was no incentive for the institutes to implement them,' she said. 'But if STIP becomes a government policy, people will take notice. All departments will have to accept it, irrespective of who the head is. Everyone will have to consider the policy guidelines while designing programmes.' She is buoyed by the possibility that the policy will encourage institutes to establish their own gender and equity offices. 'This is the first time any Indian policy has a chapter on equity and inclusion. This is fantastic!'

When Rohini said she was optimistic that STIP 2020 could be a new beginning, we wondered if this was giving her any déjà vu. Didn't the INSA report from 2004, too, seem like just the beginning? And *Lilavati's Daughters* as well? Did it not seem like the start of something good? Why is it that nearly two decades after our first steps into dealing with the gender gap in Indian science, we are still only making new beginnings?

Until things change at the policy implementation level, the burden of demanding equality for women in science continues to rest on the shoulders of women themselves. We met many women in science who have, in some way or another, become agents of change. One popular target area for individual scientists is gender diversity during scientific events. Conferences are an extremely useful medium for scientists to be noticed in their respective fields as well as to make potentially career-altering connections. Being invited as a speaker is a huge opportunity but for some reason, one that is hard to come by for most women scientists. When Shobhana Narasimhan was invited in 2017 by the ICTP to organize their 18th Total Energy Workshop in Trieste, Italy, she and her two co-organizers—also women—decided to see if they could do something about the biennial conference's historically skewed gender ratio. She said, 'We made a special effort to think of women speakers. We nominated scientists who were ignored in the past for reasons such as 'she has young children' or 'she would never agree to come and talk', but we picked them anyway'. It was then up to the conference's advisory committee to vote on and finalize the speaker list. Shobhana was pleased to see several women get chosen, eventually.

'It's not like we told them to vote for women. We just made sure many of the nominees were women. In the end, we received feedback that it was an excellent high-standard conference, one of the best so far.'

That year, the physics conference had seven women speakers (out of 23 speakers). While still nowhere close to parity, it was a significant increase from the previous editions, which never had more than four women speakers.

Shobhana is inspired by fellow scientists who walk the talk when it comes to the diversity issue—women like her who refuse to be tolerant of sexist comments and 'manels' (a popular way to describe a panel consisting of only cis males). In a public lecture, she likened these women to the insects buzzing around a big bull (the powerful male boss), saying, 'I'm not sure about change, but at least we could irritate him.' Shobhana brought up the example of a senior scientist friend abroad who boycotts conferences that do not have enough women speakers. We asked her what would happen if somebody took such a stand in India. 'I don't know if the reaction would be good,' she admitted, recalling being chastised by senior men for suggesting that all conferences require a certain percentage of women speakers in order to receive funding. Shobhana has, on occasion, been told off by the higher-ups but that doesn't stop her from calling out their attitudes. 'It's not a comfortable position to be in,' she said.

'Even now, people warn me that doing this will hurt my career. They ask me why everything is about gender for me. It's awkward because I have to continue to work with these people, even if I find some of their attitudes sexist. But now I feel I have reached a level where I don't care that much. For younger women, it is much harder.'

Rohini Godbole too has mixed feelings about her reputation as a WiS advocate. 'I'm not sure I like it too much,' she said. 'I campaign for WiS but that is secondary to my being a scientist. Somehow, the roles are reversing. I have achieved something in the realm of physics that is not insubstantial. Hence, it hurts when this aspect is downplayed and only the women's advocacy role gets emphasized.'

It is not an exaggeration to say that being a woman doing science in India can be downright unpleasant. Vaishnavi Ananthanarayanan considered herself one of the lucky ones when she was recruited as a PI into a newly formed department at the elite IISc in 2014. In just the next three years, Vaishnavi managed to turn heads with her research and accomplishments. She was selected as an European Molecular Biology Organization (EMBO) Young Investigator and won a Science and Engineering Research Board (SERB) Women Excellence Award. She was named a 'Cell Scientist to Watch' by the prestigious *Journal of Cell Science* and was also inducted as a member of its Editorial Advisory Board. It was pretty clear that this young PI was going places.

Yet, within the IISc, all was not well. Her close colleagues were amiable, but Vaishnavi was disconcerted by frequent microaggressions in the male-dominated space. For example, there was the senior male professor who would enthusiastically greet the male colleague she was next to while ignoring her completely. 'Initially, it hurt quite a bit,' said Vaishnavi, 'I wondered what I was doing wrong.' She tried to tell herself that these were small issues compared to what many other women in science in India faced. 'I even thought, "oh at least I have not been sexually harassed" . . . that's a terrible thing to say, isn't it? But it's the real situation.'

A position at the IISc is highly coveted and is usually a lifetime gig. Vaishnavi didn't think she would leave any time soon, but when a better-looking opportunity came up in Australia, she welcomed it. Such a move can never be prompted by a single factor alone but Vaishnavi has been unshy to state that her treatment as a woman in science is one of the reasons for her leaving the country.

'It was a daily struggle to be taken seriously and to do my science the way I wanted. I moved to do better science, of course, but also because I was in a place where women were not taken seriously.'

'I don't think anyone cares, though. It seems to be that the people in power are only interested in staying there by maintaining the status quo.' The status quo is something Vaishnavi made sure to give a little jolt to shortly before she left the IISc. Along with US-based neuroscientist Shruti Muralidhar, Vaishnavi founded BiasWatchIndia,[6] an initiative that 'documents gender representation and combats gender-biased panels in Indian conferences, meetings and talks'. BiasWatchIndia was supported by funds from Vaishnavi's EMBO Young Investigator grant. Their demands for representation and accountability to organizers of male-dominated scientific events were sure-footed and caused some much-needed discomfort in the complacent Indian science community. While the duo expected resistance, Vaishnavi was sometimes caught by surprise when it came from unexpected demographics. She said: 'We expect the older folks to be resistant to the idea of including a more diverse set of speakers but for us, the surprise comes when students who are part of the institute come to the defence of organizers. Sadly, the younger generation seems to be buying into the meritocracy idea that excludes specific sections of the population.'

The fact that young scientists are often opponents of equity measures is indeed disheartening. However, this makes sense when looked at through a caste lens. Though more and more male and female students are entering STEM fields each year, a vast majority are upper caste and, thereby, under-educated on the existence of historical social inequities. This manifests in the form of extreme defensiveness, stubborn anti-reservation stances and a reluctance to think critically about merit.

Occasionally, the BiasWatchIndia team has been successful in urging organizers to rethink their panel composition. But more frequently, they are met with indifference and weak claims such as 'there are not enough women in this field' and 'we invited someone but she was busy'. The subtext of such lines of defence seems to indicate that our country lacks women with expertise in STEM subjects, with underrepresentation being a natural outcome of this.

In 2017, a group of physicists conducted an experiment to demonstrate that good representation is not as hard to ensure as some make it seem. It started the previous year at a meeting of physicists planned by Prajval Shastri, an astrophysicist who studies giant black holes in distant galaxies, to discuss her proposal for a gender-in-physics working group, which would be under the aegis of the Indian Physics Association. During this meeting, the issue of the lack of women contributors in most scientific journals came up. Unwilling to sit back and let things be, they decided to attempt an all-women's authorship experiment. Three Indian journals, *Resonance*, *Physics News* and *Current Science* were picked, whose fraction of women authors ranged from a maximum of 15 per cent to 10 per cent and sometimes even lower. Prajval's proposal to take over the helm as guest editor to produce entirely women-authored issues on the occasion of International Women's Day the following year was accepted by *Resonance* and *Physics News*. With *Current Science*, the group was forced to compromise when the editorial board declined to have a whole issue authored by women. They did however approve a women-only 'special section' within the main issue. Having convinced the respective editorial boards, six of them got cracking on what they anticipated could be a challenging but fruitful mission. Prajval was joined by Sudeshna Mazumdar-Leighton for *Resonance*; Bindu Bambah and Vandana Nanal came on board for *Physics News*; and Sulabha Kulkarni and Neelima Gupte guest-edited the special section for *Current Science*.[7]

For Prajval, the intention was clear. 'My goal of having an all-women's authorship was not to 'showcase' women's work. It was to just demonstrate that it is absolutely normal and ordinary for women to write competently about physics for a variety of audiences.'

The production of the issues went fairly smoothly. Prajval, who co-edited two of the three special issues, was pleasantly surprised when nearly all of those who were invited to contribute accepted. However, when we asked Prajval to reflect on the experiment years later, she said, 'It was silo-ization from the start.' She explained that *Current Science*, unlike the other two journals, is a research journal where scientists submit research articles on their own. There was no concept of being 'invited' to do so, as there was with *Resonance* and *Physics News*, which published content for a more general audience. Nevertheless, there were still ways to make their idea work.

'The editorial board could have easily decided that for the March issue they would only consider contributed pieces from women. Or, there could have been a special edition of review articles (which are typically submitted on invitation only). The board refused to do either. They finally only permitted this silo-ized 'special section'.'

As Prajval and Sudeshna wrote in their editorial letter in *Resonance*, 'Such a "special" issue should not have been necessary.' They ended the letter with the ambitious hope that such an initiative would not be necessary again. Indeed, the success of this experiment proved that there was no paucity of competent women authors. With just an unbiased selection of authors, the regular editorial boards could easily enhance the diversity of their content henceforth. So, did that happen?

'The editorial board clearly did not read our editorial. They seemed to just think of it as a vacation in March, and came back in April to business as usual,' recalled Prajval. She, however, fondly looks back at the experience she had with the all-women production team at the IASc (which publishes *Resonance*). 'They deeply and repeatedly appreciated the effort and the experience of working with me, which was very heartening for me. They wished I were on the editorial board!'

The experiment was met with praise from all corners. Most notably, the President of the UK's Institute of Physics, Julia Higgins, wrote to her Indian counterpart, Dinakar Kanjilal, congratulating Prajval and the Indian science community who supported her on this mission.

When the president of the academy asked Prajval to repeat the performance for *Resonance* the next year, she refused. 'It was not meant to be repeated, it was a demo. I was the one who assembled the authors, and I have shown that it was the most trivial task of all the tasks I had to do for the issue. It was so easy to find a bunch of competent women scientists/writers who said yes.'

So, Prajval informed the president that the onus was now on the editorial board to simply ensure a minimum of 50 per cent women authors in every issue. 'This was ignored. Clearly, there was no conversation with the editors and the fraction of women authors went back to 15 per cent—business as usual.' Special issues 'Celebrating Women in Science' continued in the following years at *Resonance* marking Women's Day.

For far too long, the WiS movement's progress has relied on the untiring efforts of women scientists themselves. While there have been small victories along the way, this has largely been because of the presence of influential women among the upper rungs of academia and governance. These victories are of most benefit to upper-caste and cis women. Trans persons in science, scientists from Dalit, Bahujan or Adivasi backgrounds and disabled persons remain excluded because there is no one to represent them at the top. And as long as issues pertaining to them continue to be ignored, there is no chance of having more of them at the top. Just another vicious cycle, stagnating our goal of an equal science.

The voices from the caste and gender margins are getting louder. On 29 October 2021, Deepa P. Mohanan, a Dalit research scholar doing her PhD in nanotechnology at MG University in Kerala, began a hunger strike in front of her university to protest the prolonged caste discrimination she had been facing from the director of her institute. The accused professor was removed from his post a week later, in a move that was heralded as a landmark victory in our struggle against institutional casteism.[8]

As one of the very few transgender scientists to hold a faculty position at a university, Bittu K. has been using his position and privileges to improve the situation for the next generation of trans people in science. He is very willing to acknowledge the benefits that his caste and class status have given him.

'I may not want to take these privileges but I know that I do and that they operate in ways that I cannot fathom. I see how they operate against my friends who are Dalit, Bahujan or Adivasi in academia, and they are extremely severe.'

While at the University of Hyderabad (UoH), Bittu was deeply hurt by the suicide of his friend Rohith Vemula.[9] Rohith had suffered persecution and discrimination from the government and the university. He was a student of science who dreamed of becoming a science writer. The tragedy led a group of people at the university to start the Rohith Vemula Science Club. Being part of this community reaffirmed Bittu's belief in the need to democratize science.

'Many people from Dalit, Bahujan and Adivasi communities came to science for the first time [through the club] and everybody had great perspectives that opened my eyes. Earlier, they used to feel that the humanities and social sciences were the only spaces open for them to participate academically. Now, these are some of the most committed nerds I know. These are people with whom I share science news.'

At the UoH, Bittu was helped by student activists to set up a committee for trans students. The intention for Bittu was that some sort of structure existed to help potential trans students at the university deal with problems even after he left. While Bittu was not institutionally permitted to operate as the gender of his choice at UoH, Ashoka University has been more accommodating. 'When I joined, they agreed to immediately change my gender and name on record and ensured that I had access to appropriate bathrooms,' he said. Bittu is pleased to see that since then, gender-neutral bathrooms have come up in multiple spaces on campus. 'In Ashoka, we have a nice system where trans students experience a gender-affirming environment as soon as they join. We are also trying to draft a trans policy that will, hopefully, serve to be useful,' he said.

As a professor at Ashoka, Bittu has been able to reach out to several trans students who are in dire need of company. 'The journey of being a trans person in science has been very lonely,' he admitted. 'The perspective of gender and science has been largely reduced to women in science and trans scientists have very little space in this discourse.' He supports a number of trans students who are struggling and has also established an informal trans mentorship network to help people get in touch with each other. 'It has helped trans students, who are almost always the only trans person they know in the institution, feel less alone.'

The visible presence of trans scientists like Bittu at Ashoka University and A. Mani at ISI Kolkata is inspiring a slow and steady rise in discussion about transgender people in science, mostly among younger stakeholders. The mention of words like 'genders' and 'transgender' in policy interventions such as GATI and STIP 2020 suggests that the decision-makers have taken notice, but their approach seems far too tentative to expect radical change. And the status quo is so miserable that anything but radical change is really quite useless.

'Women in STEM' has become a trending topic in the 2020s, compared to when we started out on this lab-hopping journey less than a decade ago. Politicians and tech corporations have all appropriated the cause in their manifestos and speeches, as has the popular media. There have been at least two big-budget Bollywood films about women in STEM. While appreciating the progress we have made, it's also worth reminding ourselves that it's not the media, corporations or even the government, that have taken great personal risks to move us forward. It's the relatively little-known women and individuals from various margins who did this. The ones who undertook surveys, built communities, challenged authorities and staged protests to stand up for themselves and others. Every time systemic apathy grinds this movement for equity to a halt, we rely on these tremors from various pockets of Indian STEM for a jumpstart.

19

Gap Can Worsen

After hundreds of hours of interviews, audio recordings and all the resulting reports, we sit in front of our screens on the shared virtual editorial desk, many miles between the two of us. And we are more restless than ever about the 'gender gap in Indian science'. As editors, reflecting on the last seven years, we go back and forth on how to best articulate it. The problem of access to Indian laboratories affects various axes of marginalization—caste and gender are just two of them. Regardless of how we word it, we now understand that this gap can only be 'closed' with a shift in the culture of science. To make way for true inclusion, the patterns of exclusion that we have reported on must no longer be replicated in our labs. True equity can only be achieved when our STEM culture starts to chime in with its own values of openness, critical thinking and collaboration as a progressive human endeavour. On the other hand, continuing with the status quo and misdirected policies that fail to address the underlying causes of the gender gap can only make matters worse.

With the lab-hopping project, we were able to punch further openings within the ivory towers of science institutes in

India, from which the voices of marginalized people in science have been trickling. We hear these voices pooled together in the virtual communities and spaces where our project operates. As intersectionality is damming up together, we wonder, if and how the flood of change will come.

'My biggest fear is that the gap will worsen,' said Anitha Kurup, a sociologist at NIAS Bangalore and the first author of an extensive data scrounge on Indian women in science. In the previous chapter, we discussed the same study with Anitha's co-author Rohini Godbole. The study, which is now over a decade old, provides clues not only to the reasons behind the gap but also contrasts the standpoints of different women in science as well as the men included in the survey. Here are some of its important findings:[1]

- 14 per cent of women in research never married compared to only 2.5 per cent of men in research who never married.
- 86 per cent of men scientists compared to 74 per cent of women scientists reported having children
- Higher percentage of Women in Research spent 40–60 hours per week at work; higher percentage of Men In Research reported spending less than 40 hours per week at work.
- More women responded that organizational factors such as flexible timings, day care, transportation and accommodation influenced them to take up scientific positions they presently had. More men than women had left previous posts for better prospects.

In essence, what we can tell from the only large official study we have is that a woman is most likely to stay in science if she is single, childless and spends longer hours at work than most men. How many women in any strata of Indian society do you know that would fit the bill? The odds are stacked against most women. By design.

The critical part of Anitha's study was that, as a sociologist studying Indian women in science,[2] she was interested in the perceptions of her diverse study group—in other words, what was their answer to: why do women drop out of science? And indeed, when questioned about why they think women drop out, most men said that it is because of family. However, actual results when the women themselves were asked proved to be different. Family responsibilities was not the top reason cited by women participants in the study for their dropping out of the research. Instead, it appears that the majority of women who leave research in our country for one major reason—because they do not get the opportunity to stay, said 66.7 per cent of women who had applied for jobs but were jobless. 'Disabling organizational factors' that include long/inflexible hours, no room for professional growth and the lack of daycare facilities were the other reasons from the top for not taking up science positions.[3]

Clearly, your identity and lived experiences impact how you make sense of the gender gap. Anitha's (and others') fear of the gender gap worsening comes from reading two things together: what her data is telling her, and the fallacies on which many current attempts to close the gap are based. From scanning through the current schemes and government programmes, it becomes apparent that women in science are not the ones coming up with these remedies, nor are they being heard.

Programmes to close the gap seem to be fixated on getting more girls into science and benevolent service towards mothers and daughters-in-law who took a break from science to fulfil familial responsibilities. A release from October 2020 on the DST's website quotes the then head of the DST, Ashutosh

Sharma as he launched Vigyan Jyoti, a government programme inspiring girls into science:

> Considering the fact that women leadership and role models are very limited in many areas of science & technology, we need to start with the inputs to the pipeline—ambitious quality inputs that are definitely going to succeed and therefore provide the next-gen leadership and role models without worrying too much about the glass ceiling.[4]

The release referred to the programme as one meant to build confidence among young girl students. It further quotes Ashutosh as saying: 'After High school, a lot of girls have no clear idea of what they are going to do or what their ambitions are.' The government secretary seemed to be very concerned with finding ambitious girls of 'good quality' that can be 'inputted' into the science pipeline, thanks to the inspiration provided by his department. So good the quality of the inputs must be, that the department will no longer have to worry about the glass ceiling, which is commonly blamed for preventing women from reaching top positions. Such benevolent sexism dished out by government officials fires up the feminists in science. The 'fixing the women' approach is flawed because women do not need fixing. That's not where the problem is,' astronomer and gender equity crusader Prajval Shastri has had to repeatedly state.

Anitha, the first author of the aforementioned study, addresses misinterpretations of the gender gap in a more prosaic manner:

> Important headway can be made on the gender gap by addressing organizational and infrastructural facilities as

well as by undertaking policy changes that may be critical to attract and, more importantly, retain women in scientific research. Such interventions need to move beyond the traditional framework that locates societal and family duties as singular factors responsible for women dropping out of science.

There are many other feminists in Indian science who, like Anitha and Prajval, are critical of the dominant official perspective of the reasons behind the gender gap. To them, the approach to existing efforts to close the gap is condescending to all women and girls and even revealing of how the real reasons behind the gender gap—patriarchal biases, harassment and the boys' club—are skirted around by cis men in command. The entry-level problem of women in science is a more convenient problem for authorities to tackle. They only need to organize open days at research institutes and publish books on role models every year to make it seem as if they are doing their bit. In contrast, the removal of the obstacles faced by women who are already in science calls for a more brutal shake-up of institutions. Necessarily, it would require the people leading institutions to confront their own biases.

The discourse around 'women in science' has grown considerably since we began the lab-hopping project in 2016. While we are encouraged by the evolving discourse, we are also sceptical. Whenever problems faced by women in science are discussed, the focus is usually on aspects such as the demands of motherhood and domestic responsibilities. Rarely is adequate attention paid to systemic issues such as sexism in labs, male-dominated scientific meetings, sexual harassment and women-phobic policies. Consequently, women in science are only heard when they have an inspiring story to tell, or when they have achieved something. Focusing on the positives is the only accepted way to talk about women in science in India today. When one speaks about unpleasant realities and holds individuals or entities accountable for the situation, the support for women in science whittles away. The efforts to merely celebrate women's contributions to science are a pre-eminent way to discuss women in science. Relying on this approach is counterproductive as it seems to justify the status quo and legitimize the struggle to survive in science, leaving us with the hope of slow 'organic' change. The vision of open and inclusive science is relegated to the indefinite future.

To move forward, the Indian science establishment must shed the perspective that the gap can be closed by increasing women's and girls' access to education. And instead, interrogate its organizational frameworks that keep women and other marginalized groups out—this is where amends are needed. In her large study published in 2000 looking at organization features of doctoral students in the US, Mary Frank Fox, who is often called the pioneer and leader of 'women in science' studies, wrote:

> Twenty years ago, observers of trends (including feminists) projected that occupational equity for women in science was a 'matter of time' for the increasing cohorts of women Ph.D. graduates to mature and move through the ranks. However, these assumptions and beliefs that women's growing access to education would result in gender equity in scientific careers have proven unfounded.[5]

Back home, according to the UGC annual report from 2015, 44.2 per cent of science PhDs in India were women.[6] UNESCO data from the same year states that women made up only 14 per cent of the scientific workforce in the country. This large loss of women scientists at the PhD level has been dubbed 'the waterfall' by yet another feminist in Indian (nano)science, Shobhana Narasimhan, whom we met in the previous chapter. Shobhana brings up 'the waterfall' whenever she talks about Indian women in science. She does so in her efforts to challenge the narrative of the steady leaky pipeline that is more suitable to the attrition patterns seen in the USA, which does not apply to the Indian context, according to her. Shobhana's waterfall is a clear signpost for where the gender gap lies and where the action is needed.

The belief that the closing of the gender gap is only a matter of time, and organic growth is coming, is a folly often repeated in the Indian science community. We have heard mainly established officials and scientists repeat this false hope. A well-meaning cis man who supports women in science explained the gender gap in this way: 'It is likely also a function of the era—in earlier times (in fact, till a few decades ago), there were hardly any women in major positions in science where they could shape things going forward. Of course, nowadays things are different (and much better),' he said.

Often, when feminists in science deliver talks, they are armed with responses to such narratives that they know they will encounter. For instance, Shobhana follows up her waterfall metaphor with: 'You may think that it is just a function of time and that things are changing now. But sadly, that is not the case. When you trace back, you see there have been a rather large number of women in science up to PhD level for a long time.'

Besides hard stats, examples in history also show that change in this regard is not linear. The Indian Women Scientists Association, set up in the year 1972, began with the resolve to collectively voice issues of women in science that were only then becoming perceptible to India's research institutes.[7] Today, the ageing organization is only a ghost of its former self. When we visited the headquarters in New Bombay in 2019, there was no sign of any advocacy for closing the gap. More recently, editors of the book *Lilavati's Daughters* bemoaned the lack of momentum to close the gender gap; they said the momentum brought forth by the popularity of the book and the academies' efforts that led to it should have been sustained by the Indian scientific community. We have learned from our lab-hopping journeys that the 'time will close the gap' narrative is at best a misunderstanding, and at worst a distraction from the ongoing gatekeeping. Mostly, it is a strategy to assuage the unpleasant realization that the Indian science community is far from being equitable.

*

At various WiS events through our lab-hopping years, we have heard the science and technology bosses say: 'discussion is fine, but give us action points/recommendations'. The first recommendation that feminists in Indian science have been making remains: 'Start engaging with the real issues! The data is already there.' As part of the backlash from those in power today, there has been an incessant demand for data in order to 'prove' where the problem occurs and what can be done about it. Preeti Kharb, an expert on black holes in galaxies has collected annual gender data as a ritual part of her duties at Astronomical Society of India's Working Group for Gender Equity (WGGE).[8] She told us in 2020:

'Scientists claim they love to see the numbers and the hard statistics. I have been gathering these numbers for Indian astronomy research institutes for the last five years and presenting them at the start of our ASI meetings. They then approach me to 'explain' the low numbers instead of contemplating them on their own They have developed a rationale for the low numbers but that rationale never includes personal implicit biases or institutional sexism (not providing childcare facilities, devaluing the work of women researchers, etc.). Many of us, including me and Prajval, have been giving extremely 'scientific' talks where we list all the proposed reasons for the low number of women in academia and then systematically, with DATA, rule them out one by one.'

'Ideally, these talks should have completely changed the minds of the number/data-loving top-scientists and officials,' said Preeti. But still, change seems elusive. Preeti followed up her comments by suggesting: 'Therefore, the issues are complex. And perhaps we will all need to devise some complex means to address them?'

Indian feminists in science are sceptical of programmes 'inspiring' schoolgirls into science with the aim of closing the gender gap. These 'fringe' efforts, they say, are not as crucial as removing the structural and biased barriers within institutions that keep different groups out of our labs. More efforts, they believe, need to be put into training the science community in its entirety about biases holding women back. Along with this, they also asked to critically examine and rectify the different schemes instituted by the government, like the re-entry schemes that have failed hundreds of women.

Srubabati Goswami, a particle physicist at PRL and another long-time WiS crusader, echoes the need for a multipronged approach. 'In general, I think since the system and all rules are made by keeping men in mind, one has to think about what changes are needed for women,' she said.

Tailor-made policies can come to the aid of those in need now and solve a specific problem, like those who took a break from science or those who have had to relocate and leave their previous labs for any reason. Srubabati made one such recommendation for specific and immediate change at a 2019 women-in-physics conference she helped organize.[9] She brought the conference to a close by clearly listing out recommendations that had emerged. One of them was to adopt a mandate for all national conferences to invite a fixed number of women speakers. Srubabati said,

'I have seen that when people are selecting speakers for conferences and schools in India, names of women are less suggested. Many places abroad are making it a point to avoid this. The IUPAP has specific instructions for this. But in India, so far, I have not seen scientific societies, institutes, academies or departments take such a stand.'

Our science departments do make some efforts in this regard: for example, the DST's mobility scheme is designed to help women scientists who have had to relocate. But these are not gender-neutral and end up reinforcing patriarchal biases.

Gender neutrality is a critical example of policy attempts that go beyond solving specific issues but bring about a broader cultural-setting change. Even though the recent 2017 amendment to the Maternity Act mandates establishments to set up crèche facilities for 50 or more employees of any gender, it fails to broaden the scope of childcare by mandating only Maternity Leave and ignoring Paternity Leave. Further, various details are left out in the mandated law, leaving it open to interpretation. For example, the Maternity Act is silent on ensuring how to set up and maintain the quality of creches. Srubabati believes that gender neutrality in policies and how they are read by institutes can ensure a good life in science for everyone involved.

'Many times I have heard people say "the Institute has a creche for women employees". But creches are useful not only for women but also for men. These are all due to biases that are imbibed in our system. Childcare is the responsibility of all kinds of parents and society in general and by extension the employers and our institutions,' she said.

Also weighing in on policy matters is Shambhavi Naik, Head of Research at The Takshashila Institution in Bengaluru: 'Now it is very well recognized that we need more diversity. India's push for gender inclusion in science is welcome, but it has to go beyond tokenism and mere intent.'

'These conversations have to be more aggressive and robust rather than just being about changing the narrative. We need definite policy changes with predetermined objectives,' said Shambhavi, a researcher who studies public policy in India, addressing an online conference called NeuroFemIndia in 2021.[10]

> Policymaking has to be scientific, and it is key that we discuss the challenges in the Indian context. And this begins by hearing the voices of the marginalized, those who are not privileged to have a voice in the first place.

In her work of examining science policy in India for a number of years, she has observed several inadequacies:

> A drawback of a lot of our policies is that we don't have monitoring mechanisms to show if long-term objectives of inclusion have been met. Another problem is that our policies have been looking at issues in silos. For example,

if you consider the woman scientist award, we look at how many women have been awarded and have finished the fellowship, but what happens to these women afterwards? Do they remain in science? Do they continue? It's not just about retention but about helping them have a successful career of their choice. Have they been able to actually achieve that? This is the question that we need to answer.

'We can start by thinking of policies in three ways: immediate, mid and long term,' she recommended, echoing the need for a multipronged approach to inclusion policy that Preeti and Srubabati's have also suggested above.

Do our science departments, institutes and lawmakers realize that their policies must tackle more things than one at a time?

Further, Shambhavi is fully convinced that there is a long way to go before Indian policy is truly inclusive and that incremental steps are needed now.

I think it's safe to say that the opportunity cost for a woman to do science is much higher than that for a man, but it's still easier for women than for other marginalized groups.

Science, as it is today, is an occupation meant for upper-caste heterosexual cis men. And with it, come many of their narratives and rhetoric that have been internalized, sometimes even by women. These need to be taken on both via policy and cultural dialogue.

What marginalized people in science tangibly need is easy access to funding, lab space and positions to move forward in a research ecosystem where they and theirs can thrive. Our reporting shows that these have been denied to them historically and strategically. At the same time, these barriers need to be removed for the few openly transgender scientists and those from marginalized castes and tribes. These groups also require urgent support at the level of access in schools and higher education. As recently as 1990, Grace Banu, who would grow up to become possibly India's first trans Dalit software engineer, was shunned from her school classroom. Her activism shows that it continues to happen today. In all her power, Grace addressed a women-in-science conference to say: the academic regime's silence on this is 'an apartheid' for trans and Dalit people.

A. Mani, a set theory expert at ISI Kolkata and a proud trans lesbian, addresses the complex problem-solving with a simple insight. 'As long as men are the dominant gatekeepers in scientific publishing and hold the vast majority of senior research positions, the social-scientific process will continue to perpetuate entrenched gender inequities. This is a legacy of the patriarchy in science,' she said in an interview.

Mani with other feminists in science, calls for firm affirmative action at the level of positions, grants and leadership in Indian science institutes and government science departments. According to them, a panel of upper-caste old heterosexual cis men with a lot of experience in science, even with all their good intentions, will definitely fail in bringing a paradigm shift. Who is in the lead matters.

Feminist colleagues would like to see not just women or trans people in scientific power positions in India but a diverse group of people taking up the space—who are progressive and recognize the issues insightfully and are not afraid to take a stand. And this representation, they say, must be allowed to smoothly function inside institutes in a horizontal way rather than just vertically, interfacing at all levels inside the institutions, including at the student level. Bittu K., a behavioural neuroscientist you may remember from the previous chapter, has worked tirelessly to instate gender-affirming realities at Ashoka University in Delhi. He said: 'I want to see more students from marginalized communities, queer and trans identities in our labs and a culture for them where they feel safe and empowered to do their science.'

The science culture we are imagining with our feminist colleagues is a culture that has no tolerance for gatekeeping. And this involves letting go of our obsession with metrics such as exam ranks, the number of papers published, grants and awards, which we currently use to indicate scientific prowess. In previous chapters, we covered such a meritocracy and how it overlooks the requirement of social capital involved in receiving these scientific accolades, which is unaffordable by most marginalized people. These metrics are first, not a proxy for the quality of research and secondly, over-reliance on these metrics has been breeding the publish-or-perish model that has enslaved young scientists, often to the point of harassment, psychological stress and data forgery— something for which Indian science is infamous. We will have to seriously question these practices and seek better alternatives. Soumya Prasad is an ecologist who identifies as an eco-feminist. We discussed her journey of dropping out of institutional science at JNU to work on the ground in Dehradun, in a previous chapter in Part II. She believes: 'If academics is a full-time devotion, then women are not going to be part of it.'

'Women are not interested in a broken system. They take on diverse roles in life and are at the forefront of community interventions. If identification of oneself with science means separating oneself from the community into ivory towers and power ladders, not a lot of women will be joining in,' Soumya asserted.

Family and community-focused policies that value diversity could cater to everyone in the science ecosystem in a gender-neutral way. If and when women and other marginalized people begin to thrive in science, they will surely change its culture. Science is bound to become safer, more productive, more democratic, more critical and more 'science'.

We are envisioning a science culture that has moved past its epidemic of mental stress and imposter syndrome in our institutes and is moving forward with a resolve to open up to the true potential of scientific research. Many of our interviewees have attested that the days of the lone genius scientist are long gone. Along with kinder ways to motivate students and colleagues to work with each other, we need anti-caste politics permeating our labs. This can ensure the complete constitutional implementation of gender-neutral laws and reservation laws in the country. As we know from various studies in recent years, central government institutions have been in violation of allocating reserved seats at the PhD admission level as well as the faculty level. True allies must take the lead in breaking the rules until there is no active stifling of political opinions in our science institutes, as we are witnessing in the 2020s.

There is a lot of rage documented in this book. The rage of those pushed away from the sacred hallways of Indian science and the rage of those who are fighting from within. This rage is our biggest strength. As our friend and colleague Sayantan Datta once wrote: 'what makes us collectively stand in solidarity with each other is our anger towards how we have been wronged, and how others have been wronged.'[11] Our enraged hearts are exhausted from waiting; waiting for the STEM paradigm to shift. The impatient arsonist within, sometimes itches to take down the violent structures and rebuild them from scratch. Other times, we catch ourselves wondering if it would not be easier to walk away from these oppressive structures that are rotting the very core of Indian STEM.

Though promising, the limitation of the clean slate approach is that it overlooks the lives of the few people inside the institutes who are trying to change it. So, we stand behind the call by feminists in science for a paradigm shift in the perception and practice of science in India.

It only takes a walk into a science classroom or a conversation with a bright-eyed youngster to see that the passion for science is very much alive in our country. Their chances of entering the scientific ecosystem lie on the backs of the brave fighters for equity from within.

Enough about leaky pipelines and sticky floors. Time for some cleaning! Who will do it?

Notes

Part I: How It Began

Chapter 1: Introduction

1. Aashima Dogra, 'Natasha and Husnara Save the Mandarins', 21 March 2016, on thelifeofscience.com, https://thelifeofscience.com/2016/03/21/natasha-and-husnara-save-the-mandarins/.

2. Christopher Coley, Christie Gressel, and Abhijit Dhillon, *A Braided River*, (UNESCO-DST's coffee table book on India's women in science, UNESCO Office in New Delhi, 2022), https://unesdoc.unesco.org/ark:/48223/pf0000380700.locale=en.

3. Mysuru college allows women staffers to avail of period leave, https://www.deccanherald.com/state/mysuru/mysuru-college-allows-women-staffers-to-avail-of-period-leave-1090440.html.

4. Divya Arya, 'Karnataka Hijab Controversy Is Polarizing Its Classrooms', BBC News, https://www.bbc.com/news/world-asia-india-60384681.

5. 'Kerala Dalit PhD Scholar Deepa Mohanan Ends Hunger Strike, Says Her Demands Were Met', https://www.thenewsminute.com/article/kerala-dalit-phd-scholar-deepa-mohanan-ends-hunger-strike-says-her-demands-were-met-157387.

6. 'Did Work Pressure Push IIT-Gn PhD Scholar to "Quit" Life?', https://ahmedabadmirror.com/did-work-pressure-push-iit-gn-phd-scholar-to-quit-life/76902068.html.

Chapter 2: About the Book

1. See https://www.nature.com/articles/513305a.
2. 'A Guide to Gender Identity Terms', https://www.npr. org/2021/06/02/996319297/gender-identity-pronouns-expression-guide-lgbtq.

Chapter 3: The Gender Gap

1. UNESCO Institute of Statistics Factsheet on Women in Science, 2019, http://uis.unesco.org/sites/default/files/documents/fs55-women-in-science-2019-en.pdf.
2. PIB Delhi, 'Union Minister Dr Jitendra Singh Says, Government Aims to Put India "Among Top 5 in Terms of Quality of Research Outcome" by the Year 2030', Ministry of Science and Technology, accessed by Press Information Bureau, 28 July 2022, https://pib.gov. in/PressReleasePage.aspx?PRID=1845792.
3. Geetha Baali, 'Why There Are Fewer Women in Science', January 2019, https://www.downtoearth.org.in/blog/science-technology/why-there-are-fewer-women-in-science-62720.
4. Soniya Agrawal, 'Women in STEM: The Growing Numbers, Challenges and Whether It Translates into Jobs', 23 July 2021, https://theprint.in/india/education/women-in-stem-the-growing-numbers-challenges-and-whether-it-translates-into-jobs/700564/.
5. The World Bank, Gender Data Portal, Share of Graduates by Field, Female (%), 2021, https://bit.ly/worldbankgenderdata
6. All India Survey of Higher Education Reports, various years, https:// aishe.gov.in/aishe/home.
7. DST, 'Research and Development Statistics at a Glance', 2019–20, https://bit.ly/dst-rnd; 'Encouraging Participation of Women in STEM by Economic Times Grey Cell', 8 March 2022, https://economictimes. indiatimes.com/tech/information-tech/encouraging-participation-of-women-in-stem/articleshow/90080845.cms.

8. Deepthi Ravula, 'Why Promoting Gender Diversity in STEM Is Crucial', 25 August 2020, https://www.thehindubusinessline.com/opinion/why-promoting-gender-diversity-in-stem-is-crucial/article32436728.ece.

9. Vineeta Bal, 'Women Scientists in India: Nowhere Near the Glass Ceiling', *Economic and Political Weekly*, Vol. 39, No. 32 (7–13 August 2004), pp. 3647–3649, http://www.jstor.org/stable/4415389.

10. Indian National Science Academy, New Delhi, 'Science Career for Indian Women: An Examination of Indian Women's Access to and Retention in Scientific Careers', October 2004, https://bit.ly/insa-2004; Inter-Academy Panel on Women in Science in India, 'A Roadmap for Women in Science and Technology—A Vision Document', 2016, https://bit.ly/interacademy-wis-2016; R. Godbole and R. Ramaswamy, 'Women in Science and Technology in Asia', The Association of Academies and Societies of Sciences in Asia (AASSA), Country Report (India) by 2015, https://bit.ly/aassa-wis-2015; 'Evaluating and Enhancing Women's Participation in Scientific and Technological Research: The Indian Initiatives', Report of National Task Force for Women in Science, 2005, https://bit.ly/dst-taskforce-2005; IAS Annual Report, Women in Fellowship, 2006 https://www.ias.ac.in/Publications/Annual_Reports/; D. Ngila et al., 'Women's Representation in National Science Academies: An Unsettling Narrative', *S. Afr. j. sci.*, Vol. 113, Nos. 7–8, Pretoria, July/August 2017, https://bit.ly/sans-wis-2017.

11. DST, 'Directory of Extramural Research & Development Projects 2018–2019', 2021, https://dst.gov.in/sites/default/files/EM_Directory_2018_19_0.pdf.

12. DST, 'Research and Development Statistics at a Glance', 2019–20, https://bit.ly/dst-rnd.

13. B.M. Gupta, S. Kumar, and B.S. Aggarwal, 'A Comparison of Productivity of Male and Female Scientists of CSIR', *Scientometrics*, Vol. 45, 1999, 269–289, https://link.springer.com/article/10.1007/BF02458437#citeas.

14. Akanksha Swarup and Tuli Dey, 'Women in Science and Technology: An Indian Scenario', *Current Science*, Vol. 119, No. 5, 10 September 2020, https://www.currentscience.ac.in/Volumes/119/05/0744.pdf.

15. 'International Comparative Performance of India's Research Base 2009–2014', A Bibliometric Analysis, NSTMIS, DST and Elsevier, published March 2016, http://www.nstmis-dst.org/Pdfs/Elsevier. pdf.

16. Bias Watch Women Faculty Base Rates, accessed on 10 April 2022, https://biaswatchindia.com/base-rates-of-indian-women-faculty/.

17. All India Survey of Higher Education Reports, various years, https:// aishe.gov.in/aishe/home.

18. R. Godbole and R. Ramaswamy, 'Women in Science and Technology in Asia', The Association of Academies and Societies of Sciences in Asia (AASSA), Country Report (India) by 2015, https://www.ias. ac.in/public/Resources/Initiatives/Women_in_Science/AASSA.pdf.

19. IAS Annual Report, Women in Fellowship, 2006, https://www. insaindia.res.in/pdf/ws_anx_1a.pdf.

20. Bangalore Mirror Bureau, 'Bengaluru: IISc Recognized as "Institute of Eminence" by the Centre', 10 July 2018, https://bangaloremirror. indiatimes.com/bangalore/others/bengaluru-iisc-recognised-as-institute-of-eminence-by-the-centre/articleshow/64922481.cms.

21. All India Survey of Higher Education Reports, various years, https:// aishe.gov.in/aishe/home.

22. Shikha Sharma, 'Women Choose STEM, but Opt Out Later, https:// indiatogether.org/stem-gap-women; R. Godbole and R. Ramaswamy, 'Women in Science and Technology in Asia', The Association of Academies and Societies of Sciences in Asia (AASSA), Country Report (India) by 2015, https://www.ias.ac.in/public/Resources/ Initiatives/Women_in_Science/AASSA.pdf.

23. *Hindustan Times*, 'RTI Data: IITs Not Following Reservation Rules for Faculty', 30 June 2021, https://www.hindustantimes. com/cities/others/rti-data-iits-not-following-reservation-rules-for-faculty-101624994117807.html.

24. Tanya Thomas, "'I Regret the Years Spent in Academia': A Bahujan Scientist Tells His Story', December 2021, https://thelifeofscience.com/2021/12/10/i-regret-the-years-spent-in-academia/.

25. Dr Vipin P. Veetil, 'Overview of the Caste Discrimination I Faced at IIT Madras', 2021, accessed via thelifeofscience.com.

26. UGC Press Note, https://ioe.ugc.ac.in/assets/download/Press_Note_for_release-Dec.7,%202018.pdf; Tata Institute of Fundamental Research NAAC Self-Study Report, 2016, https://www.tifr.res.in/NAAC/tifrSSR.pdf.

Chapter 4: Legendary Women in Science

1. Nandita Jayaraj, 'On Marie Curie's Birth Anniversary, Let's Question the Social Evils She Battled', 8 November 2017, thewire.in; https://science.thewire.in/science/marie-curies-birth-anniversary-lets-question-social-evils-battled-exist-even-today/.

2. Abha Sur, 'Dispersed Radiance: Caste, Gender, and Modern Science in India', *Navyana*, 2011.

3. Lekshmi Resmi, Prajval Shastri, Srubabati Goswami, et al., 'Gender Status in the Indian Physics Profession and the Way Forward', AIP Conference Proceedings 2109, 050019 (2019); published online, 3 June 2019, https://aip.scitation.org/doi/pdf/10.1063/1.5110093.

4. Abha Sur, 'Dispersed Radiance: Caste, Gender, and Modern Science in India', *Navyana*, 2011.

5. Siddhant Mohan, 'Remembering Fatima Sheikh, the First Muslim Teacher Who Laid the Foundation of Dalit–Muslim Unity', 7 April 2017, TwoCircles.net, http://twocircles.net/2017apr07/407472.html; Braj Ranjan Mani, Pamela Sardar, 'A Forgotten Liberator: The Life and Struggle of Savitribai Phule'; 1 January 2008: Hari Narke, 'Dnyanajyoti Savitribai Phule—1 and 2', Savitribai Phule First Memorial Lecture, 2008, NCERT, https://www.roundtableindia.co.in/dnyanajyoti-savitribai-phule-i/.

6. Sreerup Raychaudhuri, 'The Roots and Development of Particle Physics in India', *Springer Nature*, 2021.

7. Geraldine Forbes, 'No "Science" for Lady Doctors', in *Women and Science in India: A Reader*, edited by Neelam Kumar, New Delhi: Oxford University Press, 2009.

8. Neelam Kumar, (ed.), *Women and Science in India: A Reader*, New Delhi: Oxford University Press, 2009.

9. University of London, 'Leading Women 1868–2018', 2018, https://www.london.ac.uk/about-us/history-university-london/leading-women-1868-2018.

Part II: What We Saw

Chapter 5: Women on Top

1. R. M. George, 'Calling, Conflict and Consecration: The Testament of Ida Scudder of Vellore', *Christian Journal for Global Health*, Vol. 1, No. 1 (August 2014), https://journal.cjgh.org/index.php/cjgh/article/download/10/82?inline=1.

2. 'Rotavirus Vaccine Developed in India Demonstrates Strong Efficacy', May 2013, https://bit.ly/rotavirus-statement.

3. Olga Khazan, 'Why Do Women Bully Each Other at Work?', September 2017, https://www.theatlantic.com/magazine/archive/2017/09/the-queen-bee-in-the-corner-office/534213/.

4. P. Arvate, G.W. Galilea, and I. Todescat, 'The Queen Bee: A Myth? The Effect of Top-Level Female Leadership on Subordinate Females', 2018, *The Leadership Quarterly*, https://www.sciencedirect.com/science/article/pii/S1048984317305179.

5. Sandip Roy, 'Mangalyaan's Unexpected Gift: The Glimpse of Isro's "Rocket Women"', 25 September 2014, https://www.firstpost.com/living/mangalyaans-unexpected-gift-the-glimpse-of-isros-rocket-women-1729409.html.

6. Soutik Biswas, 'India's Mars mission: Picture That Spoke 1,000 Words', 25 September 2014, https://www.bbc.com/news/world-asia-india-29357472.

7. Supriya Kashikar, 'International Women's Day', 7 March 2018, features on biospectrumindia.com, https://www.biospectrumindia.com/features/18/10565/international-womens-day-supriya-kashikar.html.

8. Biocon, 'Our Legacy', https://www.biocon.com/about-us/our-legacy/.

9. Rohini Pande and Deanna Ford, 'Gender Quotas and Female Leadership', 2011, https://bit.ly/worldbank-report-2012.

10. Christopher Coley, Christie Gressel, Abhijit Dhillon, The Braided River: The Universe of Indian Women in Science, UNESCO Office in New Delhi, 2022.

Chapter 6: The Teacher–Scientists of India

1. 'All Indian Survey of Higher Education (AISHE) Report 2019–20',

2. List of Shanti Swarup Bhatnagar Awardees, https://ssbprize.gov.in/Content/AwardeeList.aspx.

3. 'Should Publishing Journal Articles Be Mandatory for PhD Students?', https://science.thewire.in/education/research-journals-phd-indian-universities/.

4. 'New International Investigation Tackles "Fake Science" and Its Poisonous Effects', https://www.icij.org/inside-icij/2018/07/new-international-investigation-tackles-fake-science-and-its-poisonous-effects/.

5. 'Let Teachers Teach' debate, https://thewire.in/tag/let-teachers-teach.

6. PM Modi's Science Congress speech, 2019, https://www.indiatoday.in/education-today/news/story/5-things-modi-said-at-indian-science-congress-2019-new-slogan-coined-html-1422636-2019-01-03.

7.   MHRD Statistics of Dropout Rate, 2020, https://indianexpress.com/
      article/explained/in-higher-education-dropout-rates-decline-in-
      last-five-years-6261594/.

Chapter 7: Rebel or Support

1.   'V.R. Lalithambika, on Launching Humans into Space', https://
      thelifeofscience.com/2018/09/04/vr-lalithambika-launching-
      humans-space/.
2.   'Grace Banu: A Dalit-Trans Technologist Fights for a Better World',
      https://thelifeofscience.com/2020/10/02/grace-comix/.

Chapter 8: Science Spouses

1.   'Mapping the Life Trajectories of Women Scientists in India:
      Successes and Struggles Anitha Kurup and Anjali Raj', https://www.
      currentscience.ac.in/Volumes/122/02/0144.pdf.
2.   Anitha Kurup and R. Maithreyi, B. Kantharaju, and Rohini Godbole,
      'Trained Scientific Women Power: How Much Are We Losing and
      Why?', 2010, http://eprints.nias.res.in/142/.
3.   'Dual-Career Academic Couples—What Universities Need to
      Know', Michelle R. Clayman Institute for Gender Research, Stanford
      University, https://www.ndsu.edu/fileadmin/forward/documents/
      DualCareerFinal.pdf.
4.   'Scientific Values: Ethical Guidelines and Procedures', https://www.
      ias.ac.in/About_IASc/Scientific_Values:_Ethical_Guidelines_And_
      Procedures/.

Chapter 9: A Hush-Hush Culture

1.   Catherine Chung, *The Tenth Muse*, 2019.
2.   'Madras HC Slams IIT-M's Gross Irregularities in Selection of
      Professors', https://bit.ly/deccanchronicle-iitm.

Chapter 10: Reframing the Leaky Pipeline

1. 'Collaboration Overload', https://hbr.org/2016/01/collaborative-overload; 'In Collaborative Work Cultures, Women Carry More of the Weight', https://hbr.org/2018/07/in-collaborative-work-cultures-women-carry-more-of-the-weight.
2. Neelam Kumar, 'Women and Science in India: A Reader'
3. 'Triple Burden on Women in Science: A Cross-Cultural Analysis', https://www.currentscience.ac.in/Volumes/89/08/1382.pdf.
4. 'Nobel Prize: Well Done Higgs Theorists but What about the Experimenters?', https://www.theguardian.com/science/life-and-physics/2013/oct/08/nobel-higgs-boson-experimenters.
5. 'Trained Scientific Women Power: How Much Are We Losing and Why?', https://www.ias.ac.in/public/Resources/Initiatives/Women_in_Science/surveyreport_web.pdf.
6. '23% Girls Drop Out Due to Lack of Toilets in School of the Country, Reveals Study', https://bit.ly/toi-dropout-toilets.
7. Ruby Jain, Madhu Kulhar and Anita Sihag, 'Drinking Water and Toilet Facilities in Women's Colleges of Jaipur City of India 2016', http://www.ijims.com/uploads/ae7ce87b45f3d71ad3b74drinking.pdf.
8. 'Problematizing the STEM Pipeline Metaphor: Is the STEM Pipeline Metaphor Serving Our Students and the STEM Workforce?', https://onlinelibrary.wiley.com/doi/abs/10.1002/sce.21108.
9. Christopher Coley, Christie Gressel, Abhijit Dhillon, *The Braided River: The Universe of Indian Women in Science*, UNESCO Office in New Delhi, 2022.

Part III: Noticing Patterns

Chapter 11: Prejudiced Remarks and Silently Sexist Thoughts

1. 'Founders of Modern Science in India', C. N. R. Rao and Indumati Rao, https://www.ias.ac.in/Publications/e-Books/Founders_Of_Modern_Science_In_India.

2. 'Monika Lends an Ear to Tradition', https://thelifeofscience.com/2016/08/30/lending-an-ear-to-tradition-with-monika

3. *Lilavati's Daughters.*

4. 'Analysis: How Implicit Biases Hamper Women's Participation in Science', https://thewire.in/gender/implicit-bias-gender-stereotypes-women-stem-association-test-lack-of-fit.

5. '"HE" is not a gender-neutral pronoun', see https://indiabioscience.org/media/articles/Spoorthi_RB_v1.pdf.

6. Alessandro Strumia, 'The Data Doesn't Lie—Women Don't Like Physics', https://www.thetimes.co.uk/article/alessandro-strumia-the-data-doesnt-lie-women-dont-like-physics-jl0bpfd9t.

7. 'Gender Disparities in Invited Commentary Authorship in 2459 Medical Journals', https://jamanetwork.com/journals/jamanetworkopen/fullarticle/2753395.

8. Kristina Lerman, Yulin Yu, Fred Morstatter, Jay Pujara. Gendered citation patterns among the scientific elite. Proceedings of the National Academy of Sciences, 2022; 119 (40) DOI: 10.1073/pnas.2206070119.

9. 'The Production of Science: Bearing Gender, Caste and More', https://www.epw.in/journal/2017/17/review-womens-studies/production-science.html.

10. 'Learning Outcomes Based Curriculum Framework (LOCF) for Undergraduate Programme (Honours) Home Science (2020)', University Grants Commission, https://zbook.org/read/141331_locf-home-science-ugc.html.

11. 'Boys to Study Home Science? That's What the Women and Child Development Ministry Wants', https://bit.ly/ht-homescience.

12. Home Science Association of India https://www.homescienceassociationofindia.com/.

13. See https://aeon.co/essays/the-idea-that-sperm-race-to-the-egg-is-just-another-macho-myth.

14. See https://royalsocietypublishing.org/doi/10.1098/rspb.2020.0805.

15. See https://aeon.co/essays/the-idea-that-sperm-race-to-the-egg-is-just-another-macho-myth.

16. 'Running an Obstacle Race with Miles Still to Go', 7th Anna Mani Lecture by Meera Nanda, https://astron-soc.in/wgge/7th-am-lecture1.

17. Dispersed radiance by Abha Sur, referenced elsewhere.

Chapter 12: The Perfect Conditions for Sexual Harassment

1. 'Forced to Work Under Sexual Harasser for 5 Years IISc Student Reveals Truth About Institute', https://bit.ly/edex-iisc.

2. 'Address Sexual Harassment in Indian Labs, Say 165 Scientists', https://scroll.in/latest/873300/full-text-address-sexual-harassment-in-indian-labs-say-165-scientists.

3. 'Sexual Harassment of Women Climate, Culture, and Consequences in Academic Sciences, Engineering, and Medicine', https://bit.ly/nap-sexualharassment

4. INSA Ethics book, https://www.insaindia.res.in/pdf/Ethics_Book.pdf.

5. 'A First-of-its-Kind National Conference Towards Gender Equity by the Indian Physics Association: Pressing for Progress 2019', Forum on International Physics December 2019 Newsletter, https://bit.ly/apf-pfp.

6. 'Raya Sarkar's List of Academic "Predators" Is the Disruptor We Needed in Sexual Harassment Discourse', https://bit.ly/firstpost-losha.

7. 'If Utsav Kadam Is a "Future Asset", Who Am I? Asks IIT-Guwahati Rape Victim', https://bit.ly/eastmojo-iitg

8. 'Team Led by JNU Prof Develops Vaccine Candidate Against Streptococcus', https://www.republicworld.com/india-news/city-news/team-led-by-jnu-prof-develops-vaccine-candidate-against-streptococcus.html; 'Johri Comes to JNU Everyday, Just to Show That He Can: Why Students Are Stressed about #MeToo Tainted Prof One Year On', https://www.edexlive.com/campus/2019/may/11/johri-comes-to-jnu-everyday-just-to-show-that-he-can-why-students-are-stressed-about-metoo-tainte-6107.html.

9. University Grants Commission (Prevention, Prohibition, and Redressal of Sexual Harassment of Women Employees and Students in Higher Educational Institutions) Regulations, 2015, https://saksham.ugc.ac.in/Home/Downloads; Implementing PoSH in Higher Education Institutions in India, https://www.ungender.in/implementing-posh-in-colleges-in-india/.

Chapter 13: Hiring Bias

1. 'Women in Academic Science: A Changing Landscape', https://www.psychologicalscience.org/pdf/Women-Academic-Science.pdf.
2. Wendy M. Williams and Stephen J. Ceci, 'The Myth about Women in Science', https://edition.cnn.com/2015/04/13/opinions/williams-ceci-women-in-science/index.html.
3. Joan C. Williams and Jessi L. Smith, 'The Myth That Academic Science Isn't Biased Against Women', https://www.chronicle.com/article/the-myth-that-academic-science-isnt-biased-against-women/.
4. 'Top Indian Chemist Helps Make the Case for Science Windfall', ScienceMag, http://ccbb.jnu.ac.in/news_files/Rao_Science-2012-Bagla-157.PDF.
5. Asia A. Eaton, Jessica F. Saunders, Ryan K. Jacobson & Keon West, 'How Gender and Race Stereotypes Impact the Advancement of Scholars in STEM: Professors' Biased Evaluations of Physics and Biology Post-Doctoral Candidates, *Sex Roles*, 2020.
6. Corinne A. Moss-Racusin, John F. Dovidio, Victoria L. Brescoll, Mark J. Graham, and Jo Handelsman, 'Science Faculty's Subtle Gender Biases Favor Male Students', *PNAS* (2012).
7. 'The Legacy of Social Exclusion: A Correspondence Study of Job Discrimination in India', *Economic and Political Weekly*, January 2007.

Chapter 14: What about the Kids?

1. 'The Raman Wife Effect: Lively Recollections', https://bit.ly/th-raman-biography.
2. 'The Power of Parity: Advancing Women's Equality in India', https://bit.ly/mckinsey-parity.
3. 'Balancing Paid Work, Unpaid Work and Leisure', OECD Report, https://www.oecd.org/gender/balancing-paid-work-unpaid-work-and-leisure.htm.
4. Maternity Benefit Act, 1961, https://labour.gov.in/sites/default/files/TheMaternityBenefitAct1961.pdf.
5. 'Mayurika Starts a Daycare Revolution—The Life of Science', https://thelifeofscience.com/2016/09/12/mayurika-starts-a-daycare-revolution.

Chapter 15: The Old Boys' Club

1. 'Researchers Collaborate with Same-Gendered Colleagues More Often Than Expected across the Life Sciences, https://www.ncbi.nlm.nih.gov/pmc/articles/PMC6485756/.
2. 'Early Coauthorship with Top Scientists Predicts Success in Academic Careers', https://www.nature.com/articles/s41467-019-13130-4.
3. Shanti Swarup Bhatnagar, https://ssbprize.gov.in/Content/SSBBio.aspx; Swarnajayanti Fellowships, https://bit.ly/dst-swarnajayanti; Infosys Prize https://www.infosysprize.org/.
4. 'Indian scientists shocked as government scraps nearly 300 awards', https://www.nature.com/articles/d41586-022-03286-3.

Chapter 16: A Taboo around Quotas

1. 'One Woman Won a Nobel This Year. Quotas Aren't the Answer, an Official Says', https://www.nytimes.com/2021/10/12/world/nobel-prize-woman-gender-ethnicity.html.

2.   Yashica Dutt, *Coming Out as Dalit*, 2019.

3.   Joint Admission Board Report, 2012, https://bit.ly/jee-report-2012.

4.   Highlights of the Recommendations of the JAB Sub-Committee for Increasing Female Enrolment in B.Tech. in IITs, 14 June 2017, https://bit.ly/jab-report-2017.

5.   Dheeraj Sanghi, 'Women Reservation in IITs', https://dsanghi.blogspot.com/2018/06/women-reservation-in-iits.html.

6.   'Puzzled For Years, IITs Finally Find a Solution to This Problem', https://bit.ly/ndtv-iit-gendergap.

7.   'IIT-JEE Will Be Missed', https://www.forbesindia.com/article/hindsight/iitjee-will-be-missed/32286/1.

8.   'IIT Alumni Loosening Up Purse Strings for Alma Mater', https://bit.ly/et-iit-alumni.

9.   'Some Colleges Set Higher Cutoff for Girls', https://bit.ly/toi-pucollege-cutoff.

10. Ajantha Subramanian, *The Caste of Merit: Engineering Education in India*, 2019.

11. 'What the Recent Scientific Misconduct Episode at NCBS Means for Indian Science', https://thelifeofscience.com/2021/07/19/ncbs_misconduct/.

12. 'Four IIT Kanpur Professors Booked for Harassing Dalit Colleague', https://thewire.in/caste/four-iit-kanpur-professors-booked-for-harassing-dalit-colleague.

13. 'Government Workers with Autism May Go Unpaid, Despite Their Valuable Contributions to Science', https://www.popsci.com/story/science/unpaid-students-autism-science/.

14. M. Van Chandola, 'Affirmative Action in India and the United States: The Untouchable and Black Experience', https://mckinneylaw.iu.edu/iiclr/pdf/vol3p101.pdf.

Part IV: Our Science Culture Must Change

Chapter 17: Nudges from the Top

1. 'The DBT-BioCARe conclave: Women Scientists Achieving Great Heights', https://bit.ly/dbt-biocare.
2. 'All These Happened to a Women Scientist', https://www.researchgate.net/publication/317823465_All_these_happened_to_a_Women_Scientist
3. Rupali Gangopadhyay and Bidisa Das, 'DST WOS-A: The Scenario from Recipient's Perspective', https://www.currentscience.ac.in/Volumes/111/08/1307.pdf.
4. On 26 January 2023, the DST announced that the WOS-A scheme has been 'redesigned' and would now be known as the Women in Science & Engineering Postdoctoral Fellowship (WISE-PDF).
5. 'Mobility Scheme (Addressing relocation issue of Women Scientists)', http://dst.gov.in/sites/default/files/ADVERTISEMENT-Mobility.pdf.
6. 'Women in STEMM: Involve the Institutions!', https://www.currentscience.ac.in/Volumes/112/04/0671.pdf.
7. The GATI Charter (Provisional), https://dst.gov.in/sites/default/files/GATI%20Charter%20%28Provisional%29.pdf.
8. Athena SWAN handbook https://bit.ly/athena-swan-handbook
9. Science, Technology, and Innovation Policy (STIP), 2020 draft, https://dst.gov.in/sites/default/files/STIP_Doc_1.4_Dec2020.pdf.
10. '"Divyangjan" Is a Controversial Word Similar to "Mentally Ill", Says U.N. Body', https://bit.ly/th-divyangjan.
11. The Twitter thread by BiasWatchIndia in response to the offensive 'men vs women' debate, https://mobile.twitter.com/biaswatchindia/status/1340869480029818881.
12. 'They Decided to Debate Women's Leadership Skills. Then It Got Worse', https://thelifeofscience.com/2021/01/02/the-debate/.

Chapter 18: Tremors from Within

1. Ramakrishna Ramaswamy and Rohini Godbole, *Lilavati's Daughters*, 2008. Published by Indian Academy of Sciences.
2. 'Science Career for Indian Women: An Examination of Indian Women's Access to and Retention in Scientific Careers', Report, October 2004.
3. 'Women Scientists Meet in Delhi on International Women's Day', 2008, https://www.currentscience.ac.in/Volumes/95/06/0709.pdf.
4. 'Evaluating and Enhancing Women's Participation in Scientific and Technological Research: The Indian Initiatives—A report of National Task Force for Women in Science', https://www.insaindia.res.in/pdf/ws_anx_1a.pdf.
5. Anitha Kurup, R. Maithreyi, B. Kantharaju, and Rohini Godbole, 'Trained Scientific Women Power: How Much Are We Losing and Why?', 2010, http://eprints.nias.res.in/142/.
6. BiasWatchIndia https://biaswatchindia.com.
7. 'Indian Science Journals Produce March Editions Authored Entirely by Women', https://science.thewire.in/science/resonance-women-current-science/.
8. 'Kerala Dalit PhD Scholar Deepa Mohanan Ends Hunger Strike, Says Her Demands Were Met', https://bit.ly/tnm-deepa.
9. 'Remembering Rohith Vemula: On His Sixth Death Anniversary, a Look at the PhD Student's Demise and Its Aftermath', https://bit.ly/fp-rohith.

Chapter 19: Gap can Worsen

1. Anitha Kurup, R. Maithreyi, B. Kantharaju, and Rohini Godbole, 'Trained Scientific Women Power: How Much Are We Losing and Why?', 2010, http://eprints.nias.res.in/142/.
2. 'Mapping the Life Trajectories of Women Scientists in India: Successes and Struggles', https://www.currentscience.ac.in/Volumes/

122/02/0144.pdf; Changing Patterns of Work–Life Balance of Women Scientists and Engineers in India https://journals.sagepub.com/doi/abs/10.1177/09717218221075129.

3. Anitha Kurup, R. Maithreyi, B. Kantharaju, and Rohini Godbole, 'Trained Scientific Women Power: How Much Are We Losing and Why?', 2010, http://eprints.nias.res.in/142/.

4. 'Mentorship Programme for Encouraging Women to Explore Broader Opportunities in STEM Launched', https://bit.ly/Dst-mentorship.

5. 'Organizational Environments and Doctoral Degrees Awarded to Women in Science and Engineering Departments', https://bit.ly/maryfrankfox.

6. UGC Annual Report, 2015.

7. Indian Women Scientists' Association, https://iwsa.net/.

8. Working Group for Gender Equity, Astronomical Society of India, https://www.astron-soc.in/gender_equity; Gender in Physics Working Group https://www.tifr.res.in/~ipa1970/gipwg/.

9. 'The Hyderabad Charter for Gender Equity in Physics', https://bit.ly/hyd-charter.

10. NeuroFEM India, 2021, https://biaswatchindia.com/neurofemindia2021/.

11. 'Raging to Reclaim Space in Science', https://thelifeofscience.com/2020/10/10/raging_science/.

Abbreviations used in this book
(in alphabetical order)

AASSA	Association of Academies and Societies of Sciences in Asia
AIIMS	All India Institute of Medical Sciences
AISHE	All India Survey on Higher Education
ASI	Astronomical Society of India
BHU	Banaras Hindu University
CCMB	Centre for Cellular & Molecular Biology
CDRI	Central Drug Research Institute
CERN	European Organization for Nuclear Research
CES	Centre for Ecological Sciences
CMC	Christian Medical College
CSIR	Council of Scientific and Industrial Research
DAE	Department of Atomic Energy
DBT	Department of Biotechnology
DRDO	Defence Research and Development Organisation
DST	Department of Science and Technology
GATI	Gender Advancement for Transforming Institutions
HRI	Harish-Chandra Research Institute
IASc	Indian Academy of Sciences

ICAR	Indian Council of Agricultural Research
ICC	Internal Complaints Committee
ICMR	Indian Council of Medical Research
IIM	Indian Institute of Management
IISc	Indian Institute of Science
IIT	Indian Institute of Technology
IMSc	Institute of Mathematical Sciences
INI	Institute of National Importance
INSA	Indian National Science Academy
IoE	Institute of Eminence
ISRO	Indian Space Research Organisation
JAB	Joint Admission Board
JEE	Joint Entrance Examination
NASI	National Academy of Sciences
NIAS	National Institute of Advanced Studies
NIH	National Institutes of Health
NII	National Institute of Immunology
PCR	Polymerase Chain Reaction
PI	Principal Investigator
PRL	Physical Research Laboratory
PU	Pre University
QIC	Quantum Information and Computation
R&D	Research and development
SSB Prize	Shanti Swarup Bhatnagar Prize for Science and Technology
STEM	Science, technology, engineering, and mathematics
STIP	Science, Technology, and Innovation Policy
THSTI	Translational Health Science and Technology Institute
TIFR	Tata Institute of Fundamental Research
UGC	University Grants Commission

UNESCO	United Nations Educational, Scientific and Cultural Organization
VC	Vice-Chancellor
WGGE	Working Group for Gender Equity
WHO	World Health Organization
WiS	Women in Science

Acknowledgements

We started thinking about this book in 2017, which, as we write this in 2022, feels like an aeon ago. Completing the book took us over three years, a period that witnessed several life-changing events for both of us, not to mention a global pandemic.

Needless to say, we could not have done it without support, starting with each others'. Special thanks to Mrinal Shah, who helped us with research, statistics and her feedback during the early drafts of this book; to Sayantan Datta who added new dimensions to our understanding of the intersectional nature of Indian science's gender gap.

Thank you to the hundreds of students, scholars and scientists who opened up to us about their journeys and trusted us with their stories. Also, to the sociologists and historians whose work on science and society helped us put everything we saw and heard in context.

To Shalini Mahadev whose nuanced thoughts and opinions about science structures and policies helped make this book more meaningful.

We would also like to thank Kiran Mazumdar Shaw for offering us book writing grants that helped us get through this journey. Similarly, the crowdfunders of our independent project also pitched in, and for this we are deeply grateful.

Thanks to Manasi Subramaniam who was the first to spot us and present us with an opportunity to turn our project into a book for Penguin. We benefited greatly from the editorial guidance from Shubhi, Shreya, Binita, Aparna and others. It means a lot to have Angela Saini's blessing for this book. Thank you, Angela, for the foreword and for the truly heartwarming conversation.

Our families and friends stood by us patiently over the years, providing soothing words when our nerves were frayed, gentle nudges when we were stuck in a rut, warm meals when we were holed up in our studies and, of course, plenty of distractions—some of them much needed.

From Nandita: Mervin and Nitin, I couldn't have done this without your constant encouragement. Amma, Acha and Mai, I'm grateful for you more than you know. Archana and Swathi, thanks for sticking with me. Fighting! ;)

From Aashima: Thank you Aamin for being patient, as well as Maa, Appa, Aarti for the support. I owe a lot to the many women and queer identifying mentors I've had. This one's for you!